SOCCER
TRAINING SESSIONS
ITALIAN STYLE

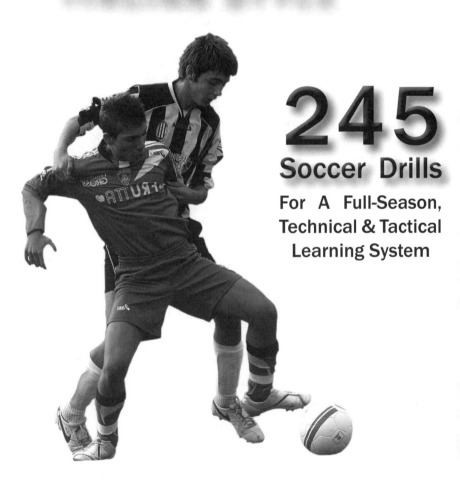

245
Soccer Drills

For A Full-Season, Technical & Tactical Learning System

Mazzantini ♦ Bombardieri

soccer learning systems

A Soccer Learning Systems publication

First published in North America in 2010 by;
Soccer Learning Systems Inc.
625 Pinnacle Place
Livermore CA 94550 USA

Printed & bound in the USA

ISBN 978-0-9844467-0-4

Library of Congress Control Number: 2010921670

Pictures: Gianni Nucci - Official photographer Empoli FC

Soccer Learning Systems policy is to use papers that are
renewable & recyclable, natural & made from wood grown
in sustainable forests.

Contents

SOCCER
TRAINING SESSIONS
ITALIAN STYLE

This book focuses on the principal characteristics of players between the age of 13 and 15, a delicate moment in the psychological and physical growth of the player.

The objective of this book is to share our training methodology, which focuses on developing technical skills and at the same time devoting the adequate time to teach the fundamentals of individual tactics and group tactics.

It is important at this age to start working on developing the tactical thought, which will improve and increase during the player's growth.

The book will also focus on motor and coordination skills and the fundamental aspects of conditioning.

The objective of each lesson plan is to develop...
- Technical aspect by way of a initial warm up game and or an activity
- Physical aspect by way of a coordination or conditional activity
- Individual tactical aspect by way of a game situation or games with a theme

The book will also introduce the first concepts of group tactics around the 4-4-2 system of play.

Mirko Mazzantini

Fausto Garcea

Tommaso Tanini

Simone Bombardieri

13-15 Year Old Soccer Players

Physical & Psychological Characteristics

Elements of Training Methodology

By Fausto Garcea

Body growth & maturation

Conditional skills

Coordination skills

Psychology for the 13 & 15 years old

Body Growth & Maturation

It is important for the coach of this age level to understand the mutation of the body the players go through.

Mr. Goding and Mr. Stratz have studied this mutation and developed three fundamental principals on which the body maturation is based.

Stratz states there are alternating phases between weight and height.

The alternating periods start with the "turgor primis" stage, from two to four years old, in which the body gains weight and the body structure assumes a more round shape.

The second phase is the "proceritas prima" five to seven years old, with a increase in height.

The third phase, from eight to ten years old in the girls and until eleven in the boys, is called "turgor secundus" which is characterized by another increase in body weight and increase in power, muscle tone and general coordination.

The fourth phase, which is the phase we are focusing on, from eleven to thirteen, is called "proceritas secunda' and it is marked by a general transformation of the size and characteristics of the player. In this period we note the increase in the length of the limbs, in the height and in the weight with loss of coordination.

The psychological aspect is also affected with the change of voice and appearance of pubic hairs in the boys and growth of breasts in the girls.

The fifth phase "turgor tertius" from fourteen to seventeen years old in the boys and from thirteen to sixteen in the girls is marked by an increase in the hormone activity and a slow down in the growth of the bone structure.

The sixth phase "proceritas tertia' from seventeen to twenty-one in the boys and from fifteen to eighteen on the girls and finally the seventh phase "turgus quartus" from nineteen to twenty-one conclude the physical growth of the individual.

Godin instead states there are shorter periods in which the body alternates phases of growth in length and phases of growth in width of the long bones such as the legs and arms. Therefore a limb undergo periods in which its length is increased and periods in which its diameter is growing.

Mr. Godin also states that in the two semesters pre-puberty the longing

of the limbs coincide with an increase in height of the individual, and in the two semesters post-puberty we witness and increase in weight. It is also important to note in the post-puberty period the muscle tone has a relevant increase and in the pre-puberty period there is an important increase in the length of the leg bones.

Conditional Skills

Strength is the ability to contrast and/or win an external resistance.

The resistance can be from your own body (jumping), from a part of the body (raise your leg), or from an external load (push and opponents).

Many factors affect the strength of an individual:

Muscle volume

Strength is proportional to the volume of the muscle, the larger the muscle the larger the strength.

Neuromuscular quality of the muscle fibers

There are red and white fibers. The red fibers have more resistant characteristics and stabilize the movements. The white fibers instead contract faster and with more intensity.

Genetic factors

Hereditary factors associated with ethnic background have an effect on the strength of an individual.

Gender

Until puberty with the same rate of training, force is identical in males and females. After puberty the force of a male and a female differs due to the different characteristics of the genders.

What we must force us know that we are dealing with sports?

There are two types of strength: absolute and relative strength.

Absolute strength is the strength an athlete is able to express regardless of its weight

Relative strength is the relationship between the absolute strength and the weight of the individual.

Who has more absolute strength and who has more relative strength between two athletes who can lift 100 kg and weigh 70 kg and 90 kg?

From a mechanical point of view, when a force is applied there is the movement, acceleration (a) of a mass (m): therefore the intensity (f) of the force to apply is the product of the mass time and acceleration ($f = ma$).

There are three types of muscle strength applied to a sport gesture: maximal, fast, resistant.

Maximal Strength or Pure Strength

It is the maximum tension developed by a voluntary muscle contraction to win a resistance.

It depends on the volume, the quantity of fibers that make the muscle mass.

Maximal strength can be trained and requires a significant effort but it should not be trained before the compete development of the skeleton-muscle system. Not to be trained before 16 or 17 years old.

Speed strength or power

It is the ability to develop strength of high intensity in the shortest time frame.

Strength endurance

It is the ability to sustain a workload of relative strength for a certain period of time.

Training strength in this age level

Depending on the sport and the type of strength training sessions are specific and designed with the individual in mind.

This ability if trained can be improved significantly. On the other hand if the muscle is not trained it will lose muscle tone and the strength itself.

As discussed earlier, until the full sexual maturation (approximately 14 years old) we should not talk about training maximal strength with weights and specific weight machines, because the intervention would be negative on the delicate and unstable bone structures.

Moreover from a psychological point of view a boring and monotonous type of training like weight lifting is not adequate for an age always in search of motivation and with a low rate of concentration.

Activities to train this ability are fun circuit training, motivational and stimulating activity alternating technical gestures and motor gestures, exercises with natural weight loads.

If we want to train the athlete through weight lifting, it is important to teach proper technique, the positioning of the lower limbs & good posture.

It is very important to exercise care if the athlete is subject to weight lifting before 14/15 year old, unless the individual is blessed with a reasonable general condition and a decent abdominal, dorsal, pectoral muscle mass.

However even if the athlete is blessed with a good physical structure it is important to follow the fundamental rules during the weight activities:

*Warm up before the work

*Never lift excess weights that only allow one lift

*Never hold a weight for a long time

*Teach the proper posture and breathing

*Always perform stretching after the activity

The type of strength mostly trained at the age of 14 and 15 is the speed strength and the strength endurance.

Speed strength

This is trained with activities at natural weight load, with constant movements at adequate speed. The repetitions should not be higher than 8/10 and the series between 3 and 5.

This is because the workload must be sharp; the muscle fibers should not tire.

Speed strength means speed and power of the muscle. Plyometrics is a methodology that uses the elastic strength accumulated by the muscle after a contraction.

Strength endurance

In soccer it is important to maintain an adequate level of strength even when the muscle starts to tire.

Strength endurance is trained with slow repetitions where significant resistances are opposed for a longer period of time.

The progression of the activity is characterized by an increment of the time the activity is performed.

The number of repetitions can be 18/20 and the loads should be between 30 and 60% of the maximum. Recovery between the 3/5 series cannot be lower than 45 seconds.

The speed of execution of the exercises must be moderate.

Circuit training with stations with the introduction of the ball and the technical gesture is probably the best type of training to stimulate concentration and enjoyment.

Endurance

It is the motor conditional skills that allow the player to extend the workload despite getting tired.

Endurance is connected to the quantity of red muscle fibers, to the functionality of the cardio-circulatory and respiratory systems, that generate energy to sustain the aerobic, anaerobic a-lactic and anaerobic lactic workload.

To train endurance we need to wait the full development of those systems which happens at around 14 years old.

Endurance can be general and specific.

General endurance does not depend on the type of activity but depends on the aerobic capacity and the capillarity or the capacity of the blood cells to supply oxygen to the muscles.

Specific endurance is specific to the activity performed by the individual.

The time of exercise should not be excessive because the exercises must be performed at a medium intensity. Recovery times are proportional to the workload.

To train this ability we can opt for mixed activities with or without the ball, circuits, relays. Ball possession, situation games, small sided games are to be preferred to runs and laps around the field.

In the event the field is not playable we can adopt a Fartlek* type of run, with change of speed and rhythm.

Fartlek, which means "speed play" in Swedish, is a form of conditioning which puts stress mainly on the aerobic energy system due to the continuous nature of the exercise. The difference between this type of training and continuous training is that the intensity or speed of the exercise varies, meaning that aerobic and anaerobic systems can be put under stress. Most fartlek sessions last a minimum of 45 minutes and can vary from aerobic walking to anaerobic sprinting. Fartlek training is generally associated with running, but can include almost any kind of exercise.

Speed

Speed represents the relationship between space and the time needed to travel. Therefore if two athletes run a 100 meters race, the fastest is the one who finish first.

In soccer speed is a very important conditional skill, but this must be combined with control of the ball. A soccer player must have a good degree of basic speed but he must be first of all reactive and explosive, with a great sense of timing and powerful in the short distances with a great ability to accelerate at the right moment with a change of pace.

A soccer player must also be quick and comfortable while running with the ball, a fundamental technical skill.

It is evident that some time the methodology of training a track and field athlete is used to train speed in soccer not understanding there is a pure speed without the ball, a motor conditional skill, and a speed while in possession of the ball, a technical ability.

The pure speed is affected by the genetic of an individual and the quantity of white fibers in muscles of an athlete. With proper training we can intervene on the efficiency of the nervous system and on the functionality of the muscle system.

In order to plan our training sessions we must consider that speed start to develop early, between the first and the sixth year, with the possibility to transform the slow red muscle fibers in fast white fibers, between 13 and 15 years old before puberty, even if in small proportions.

In soccer we must train to improve the speed of reaction, the sprinting abilities, the acceleration and the integration of these three elements that together form the speed of covering the distance required in the least amount of time.

For this perusal the psychokinetic is a methodology used a lot in soccer.

All the exercises that involve a stimulus and a reaction (various commands, various type of feedback: audio, visual etc) if they are properly planned, provided and performed can improve in the shortest time the quality of reaction to the stimulus and the kinetic reaction directly proportional to the functionality of the nervous system.

From 12 to 15 years old the acceleration can be trained and improved significantly.

The use of the ball in this type of training is important as well as the "dry" exercises (without the ball) can be useful. For example a game between two players who must sprint and accelerate for 15 meters after a visual or audio command with final shooting in goal when the player reaches the ball, are a fine example of what we mean.

It is obvious that while training speed it is important to train the running technique and the strengthening of the energetic reserves like creatine phosphate.

When the strength speed and the general mobility is trained and improved we also intervene in a positive way on the ability to perform motor acts faster and quicker. The principle for the specific training of pure speed is to plan exercises of short time, maximum 10 seconds at the maximum speed, with medium/long recovery times to allow the muscle and the brain to restore its freshness and sharpness.

Quickness

Speed is not synonymous for quickness. This skill represents the ability to perform one or more motor gestures in the shortest time.

A guitarist moves his finger quickly while playing a "virtuoso" song, just like a striker can be quick in performing combined gestures of dribbling and shooting even if he does not possess high speed base.

We need to be careful when we plan session thinking that speed is the athletic objective and instead the exercises work on quickness, or vice versa.

Agility and mobility

In the last 10 years literature considers this ability between conditional and coordination skills.

Agility represents the ability to perform gestures with the maximum agility and flexibility.

The structural properties and physiological, the joint of the newborn and children are more flexible then the adults. Cartilage rather than bones, greater ligament flexibility and low muscle tone, are factors in favor of the younger individuals.

The age we are working with pose a challenge since the athletes undergo an increase in muscle tone and in the strength limiting the flexibility and agility of the player.

Therefore it is important that in training we devote some time to the flexibility and general mobility through three methodologies: active exercise, passive exercise and stretching. Certainly the first method is the one to favor with dynamic exercises, however even stretching has its relevance and must be included in the training schedule of the coach and trainer.

Coordination Skills

As already discussed between the age of 5 and 12 coordination skills develop at a faster rate. In the age we are analyzing, between 13 and 15 coordination skills are subject to a slowdown in the development and need to adjust to compensate for the increase in body length and the proportion between the body trunk and the arms and the body trunk and the legs.

Dexterity for example is the coordination ability that by 12 years has completely developed.

Coordination skills are usually divided in general and special.

General coordination skills:

*Skill of motor learning.

*Skill of motor control.

*Skill of transformation and adaptation.

The first skill allows the acquisition of new gestures that stabilize through correction and repetition of the gesture.

The second skill allows controlling the gesture and movement depending on the target.

The third skill allows transforming and adapting the movements to the different situations that evolve from time to time.

Special coordination skills:

*Dexterity.

*Skill of combination allowing to combine various gestures together.

*Reaction skill.

*Skill of dynamic differentiation, allowing the analysis of data originating from outside and adequate the motor reaction.

*Skill of temporal differentiation allowing to give order of sequence to the motor process. In soccer the study of this skill is very important to learn the trajectory, the distance and the direction and speed of the ball.

*Skill of motor anticipation.

*Skill of motor creativity.

*Skill of orientation allowing to use properly the space.

*Skill of rhythmization allowing to express a rhythm in the motor gestures (for example the gesture of a goalkeeper that prepares himself to catch a high ball).

Balance

The balance as a coordination skill and its development and training deserve a particular attention.

Balance is the skill to maintain stability of the body during the various situations both static and dynamic.

The relevance of balance in soccer is evident. In fact a player that is able to restore quickly his balance after a tackle, after heading the ball or after shooting the ball, has more opportunities to manage effectively the subsequent movements and technical skills.

Balance is directly related to the centre of gravity. Generally a body is in balance when the imaginary line from the center of gravity to the ground falls inside the standing base which is the space between the feet.

On the contrary if that line falls outside the base the body is not in balance. This is why it is evident that a player with a low centre of gravity has more balance when moving with the ball.

Balance can be static, when the individual does not move and the centre of gravity does not move. Balance can be dynamic when the centre of gravity changes continually with movement. Balance can be in the air when the athlete is not with his feet on the ground.

Training coordination skills and the balance, it is necessary to plan and prepare activities full of creativity. Exercises used in other sports are the foundation for a complete development of a wide motor memory.

Circuit training, circuits requiring agility and the ability to change from one position to another, from a type of balance to another are necessary for an effective training plan.

Psychology For 13-15 Year Old's

According to Freud this age is characterized by the sexual organization and sublimation modifying the natural expression of an impulse or instinct (especially a sexual one) to one that is socially acceptable.

Males show more signs of aggressiveness. They favor games with rules and social games, therefore soccer is the answer to this need.

There is a valorization of "myself", the individual form his own personality and value his own feelings; at this age there is the development of a proper identity.

Parents and teachers now are criticized. Their defects are magnified and the coach can become a reference point.

In this stage there is also the formation of intimacy. Therefore we need to pay attention because even if the coach is well respected and followed, his honesty of thought and his behavior is continually tested and analyzed from the adolescent.

From a social aspect, these years are important because the individuals find mental support with each other. Therefore the soccer team has a sense of belonging, acceptance and integration.

The 13 and 15 years old finds great support in the team, and use the group as a mean to get feedback and experiment their self expression.

Therefore the sport of soccer for the individual at this age has a big impact for the rest of his life, for a social aspect, for the performance in school, for the overall health of the individual and his involvement with other activities.

Dropout from the sport is a draw back at this age and often the cause is the bad relationship with the coach and the lack of the fun element in training sessions and games.

Therefore it is important that the coach improvise himself also like psychologist other than a good teacher of the technical and tactical fundamentals.

SOCCER
TRAINING SESSIONS
ITALIAN STYLE

245
Soccer Drills

**For A Full-Season,
Technical & Tactical
Learning System**

Training Sessions

The 45 lesson plans are organized in weekly learning blocks based on three training sessions per week.

In order to allow the players a more efficient learning session, it is suggested that each learning block should be proposed for at least two weeks, except in cases where the coach wishes to work on the same concept after the regular season or over two seasons.

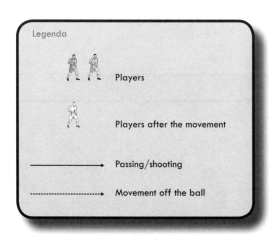

Legenda

Players

Players after the movement

Passing/shooting

Movement off the ball

Objectives

Individual tactics:
Front man marking, feint and dribbling

Group tactics:

Technical training:
Running with the ball

Physical conditioning:
Aerobic power

Organization

105

10
Warm up
Running with
the ball

15
Physical
Aerobic power

20
Technical skill
Dribbling

20
Game situation
Frontal 1 v 1

20
Game with
a theme:
Game with 6 goals and
fixed markings

20
Game:
2 fields, 5 v 5
tournament

Warm up   **10**

Running with the ball
One player with the ball is dribbling while
another player is applying passive
pressure from a frontal position.
The players change every minute.

Variation:
Dribble in a slalom pattern.
Introduce the concept of strong side
and weak side.

Physical   **15**

Aerobic power
5:00 – Fartlek run
 :40 slow jog :20 fast run

2:00 – Recovery

5:00 – Fartlek run
 :40 slow jog :20 fast run

3:00 – Stretching.

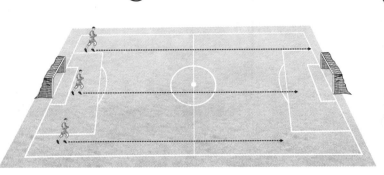

20

Dribbling

Two players with a ball each dribbles through the cones.
At the marker they make a feint and move the ball on the left side and then continue running with the ball to the end of the line switching places.

Variation:
Use different feints.

20

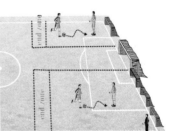

Frontal 1v1

<u>Situation A</u>
The attacker tries to score in goal. If the defender wins the ball he must dribble to the end zone

<u>Situation B</u>
Each goal is assigned a color.
The coach will call the color of the goal the attacker must score in.
If the defender wins the ball he must dribble to the end zone.

20

Game with 6 goals and fixed markings

5 v 5 game with 3 goals on each side of the field.
Goals are scored dribbling through the goals.
Each player is assigned one opponent and they must mark each other.

20

5 v 5 tournament on 2 fields

Objectives

Individual tactics:
  Frontal man marking, feint and dribbling

Group tactics:

Technical training:
  Running with the ball

Physical conditioning:
  Strengthening and explosive power

Organization

105

(20) Warm up
Warm up

(15) Physical
Strengthening and speed power

(15) Technical skill
Dribbling, feint and shooting

(20) Game situation
1 v 2 frontal marking

(20) Game with a theme:
8 v 8 with zones

(15) Game

Warm up　**(20)**

Warm up
  Group A:
  Dribbling in traffic and stretching.

  Group B:
  Coordination – various types of run, focusing on running backwards and in a slalom pattern. A typical run of a player marking and jockeying.

Variation:
Change the type of dribbling every 10:00

Physical　**(15)**

Strengthening and speed power
  3:00 15 squats maintaining the position for :10 and recovery for :10

  2:00 Stretching

  3:00 Squat, maintaining the position for :10, jump and recover for :10

  6:00 In pairs, 3 jumps over the obstacles completing with heading the ball.

15 ───────────────────────── Technical skill

Dribbling, feint and shooting

Players are divided in 3 teams. A player from each team starts by dribbling to one cone and perform a type of predetermined feint, then he dribbles to the next cone and perform another feint. The player completes by shooting in the small net.
The team that scores more goals in a set time wins.

20 ───────────────────────── Game situation

1 v 2 frontal marking (focusing on the principle of delaying)

Situation A
The coach passes the ball to the attacker who tries to attack on of the goal. 2 defenders apply pressure as soon as the ball is passed.

Situation B
Same situation as in A but the defenders starts from the position shown in the picture.

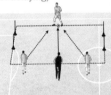

20 ───────────────────────── Game with a theme

8 v 8 with zones

The field is divided in three zones. 2 v 2 in the attacking zone and 3 v 3 in the midfield zone. Players mark only the assigned man. Goals are scored by dribbling through the end zones, and the ball must pass through all the zones. If the defender passes to the attacker they must pass back to the midfielders. If the attacker wins the ball in the attacking zone he must pass back to the midfielders.
If the midfielders win the ball they must pass back to the defenders.

15 ───────────────────────── Game

Objectives

Individual tactics:
Frontal man marking, crossing

Group tactics:
Shifting and using width

Technical training:
Running with the ball

Physical conditioning:
Quickness

Organization

(105)

(15)
Warm up:
Warm up

(15)
Physical:
Quickness circuit
with technique

(20)
Game situation:
1 v 1

(20)
Technical skill:
Defense shifting
+ 1 v 1

(15)
Game with
a theme:
8 v 8 with a lateral
v 1

(20)
Game

Warm up — (15)

Warm up
Players are divided in two teams. Each
player with a ball dribbles in traffic.
At the coach's command (visual or
audio), the defending team leaves
the ball and defend, the attacking
team tries to score by dribbling
through the small goals.
Players are assigned fixed
markings.

Variation:
Change the type of dribbling.

Physical — (15)

Quickness circuit with technique performed under condition of quickness
10m sprint + slalom around tight poles
+ 10m sprint towards a color called by
the coach. 10 repetitions

3:00 stretching
5:00 of technique in pairs
performed under condition of
quickness.

One circuit every 6/7 players

10m
10m

△ Blue
△ Yellow
▲ Red

(20) ————————————————— Game situation

1 v 1

Divide the field in 5 mini fields as shown in the picture.

Players perform a 1 v 1 with the defender applying passive pressure and moving backwards until the attacker is at the end of the field.

Ask the players to change position so the angle of approach of the defender changes as well.

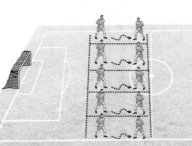
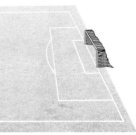

(20) ————————————————— Technical skill

Defence shifting + 1 v 1

Starting with 4 defenders located in the 4 zones as shown in the picture.

One player with the ball run across the field and the defenders perform the right movement of sliding across the field providing support and closing down when the attacker is in their zone.
Once the player has traveled across the field, 4 players start a 1v1 with the 4 defenders, trying to score in the small goals.

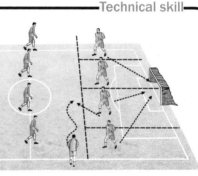

(15) ————————————————— Game with a theme

8 v 8 with a lateral v 1

Game with lateral channels. Players play a two touches game. One player of the team in possession of the ball can move into the lateral channel and if he receives the ball one player of the opposite team can enter the channel to defend.
If the attacking player wins the 1v1 he can cross without anymore pressure from the defender, unless he gets out of the channel and another player moves into the channel.

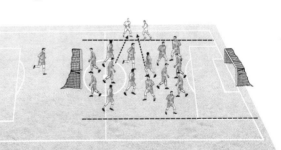

(20) ————————————————— Game

Learning Block 2   Lesson 4

Objectives

Individual tactics:
Man marking from behind

Group tactics:

Technical training:
Running with the ball

Physical conditioning:
Aerobic power

Organization

120

15
Warm up:
One touch game

20
Physical:
Aerobic power

15
Technical skill:
Dribbling in traffic
with obstacles

20
Game situation:
Man marking from
behind, ball   passed
on the ground

30
Game with
a theme:
5 v 5

20
Game

Warm up   (15)

One touch game

In a field as shown in the picture, the
players pass the ball with one touch to
the player of a different color.
When the coach raises his arm, the
player in possession of the ball
performs a "give and go" with a
player of the same color and then
perform a long pass to a player
of the different color who has
gone out of the field with the
rest of the team.

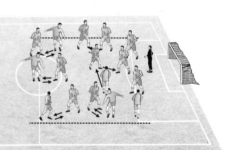

Physical   (20)

Aerobic power
7:00 Fartlek run
– :40 slow jog :20 fast run

2:00
Recovery

7:00 Fartlek run
– :40 slow jog :20 fast run

3:00 Stretching.

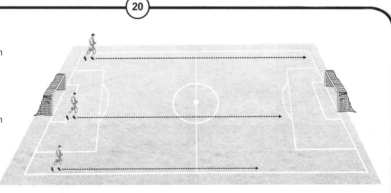

(15) ─────────────────────────────── Technical skill

Dribbling in traffic with obstacles
In a field as shown in the picture we place various objects (cones, poles, hurdles). Players dribble in traffic and working around the obstacles, first without condition then based on the coach's command (i.e. dribbling using only outside of foot).

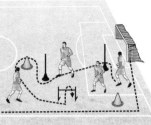

(20) ─────────────────────────────── Game situation

Man Marking from behind with ball passed on the ground
In zone A the defender marks the attacker from behind, trying to force the attacker on his weak foot.
In zone B, the defender must take into consideration his positioning from a lateral side forcing the attacker towards the center of the field.
In zone A the objective of the attacker is to score in goal.
In Zone B the objective is to cross the ball inside

the 2 goals as shown in the picture. The defenders starting position is right behind the attacker.

(30) ─────────────────────────────── Game with a theme

5 v 5 with target player and pressuring defenders
5 v 5 game with two target players. A defender stands beside each goal, and ever time the ball is played to the target player he must apply pressure and prevent the target player to turn towards the goal.

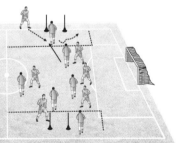

(20) ─────────────────────────────── Game

Learning Block 2   Lesson 5

Objectives

Individual tactics:
Man marking from behind

Group tactics:

Technical training:
Running with the ball, shooting

Physical conditioning:
Strengthening and explosive power

Organization

115

20
Warm up:
Ball possession and
long pass

15
Physical
Strength and
explosive power

20
Technical skill:
Receiving & running
with the ball

20
Game situation:
Individual tactics

20
Game with
a theme:
Man marking

20
Game

Warm up   **20**

Ball possession and long pass

Three teams of 5 players each.
Divide the field in three zones as in
the picture.
Team with the ball makes a long
pass to team blue.
This team must make 5
passes before making a long
pass to the yellow team.
The white team must
apply pressure in the
appropriate zone.
If the defending team
wins the ball it must
pass the ball to the
other team and
the team loosing possession must apply pressure.
(4:00 possession – 5:00 dynamic stretching – 4:00 possession – 5:00 stretching).

4:00 Ball possession
5:00 General mobility
4:00 Ball possession
5:00 Stretching

Physical   **15**

Strengthening and explosive power
4:00
20 squats maintaining position for :10
and recover for :10

2:00
Stretching

4:00
20 squats maintaining
position for :10, jump &
recovery for :10

2 jumps with heading movement
Ball from the goalkeeper
Receiving and shooting

5:00
Specific exercise as in picture 2 jumps over the hurdles, receiving ball from goalkeeper and
shooting.

(20) ───────────────────────────── Technical skill

Receiving and running with the ball – shoulders to the goal

Three players standing in the three squares as in the picture, receive the ball from the teammate and turn towards one of the goals called by the coach. They must dribble through the goal while the teammate who passed the ball applies pressure from behind.

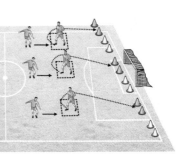

(20) ───────────────────────────── Game situation

Individual tactics: man marking from behind with ball passed on the ground

In zone A the defender marks the attacker from behind, trying to force the attacker on his weak foot. In zone B, the defender must take into consideration his positioning from a lateral side forcing the attacker towards the center of the field. In zone A the objective of the attacker is to score in goal. In Zone B the objective is to cross the ball inside the 2 goals as shown in the picture.

Variation:
Defenders start from a distance of 4 meters from the attacker

(20) ───────────────────────────── Game with a theme

Man marking

The field is divided in a central zone and in 6 mini fields. One attacker and one defender in each mini field with the attacker with the shoulders to the goal. After 4 passes in the central zone the ball can be passed to one of the attacker. If the attacker beats the defender he can shoot in goal without pressure from the defender.

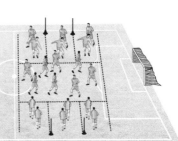

Variation:
One midfielder can double team with one defender.

(20) ───────────────────────────── Game

Learning Block 2   Lesson 6

Objectives

Individual tactics:
Man marking from behind, ball
passed on the ground

Group tactics:
Shifting and support

Technical training:
Running with the ball

Physical conditioning:
Psychokinetic and quickness

Organization

110

15
Warm up:
Exercise on
providing support

15
Physical:
Quickness circuit
with technique

20
Technical skill:
Two situations of
1 v 1

20
Game situation:
Shifting of the
defensive unit +
4 v 4

20
Game with
a theme:
Game with 4 color

20
Game

Warm up   15

Exercise on providing support:
Tactical exercise for the defensive unit
focusing on providing support. The
coach calls a color and the player
standing in the zone of the color
must quickly close down the
space, while the other players
must provide coverage.

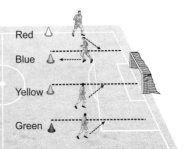

Physical   15

Quickness circuit with technique performed under condition of quickness
7:00 – lateral run in the circles, 10m
sprint, lateral run in the circles, 10m
skips with high knees and high
heels, moving left for 5m and right
for 5m, receiving the ball and
running with it at high speed.

3:00 stretching

5:00 technique in pair
under condition of
quickness.

10 mt   5 mt   5 mt   15 mt

32

(20) ──────────── Technical skill ──

1 v 1, attacker with shoulders to the goal and defender applying pressure

Situation A: 1 v 1 after a coordination
exercise with circles and cone
Situation B:
1 v 1 with the players starting in a
small square. The attacking player
makes a movement of checking
out to go receive the ball with
his shoulder to the goal.

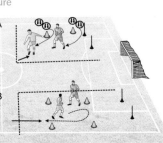

(20) ──────────── Game situation ──

Shifting of the defensive unit + 4 v 4

One player starts dribbling across
the field in front of the back 4. The
defenders work on closing down and
providing support to each other in
relation to where the ball is. At the
end of the dribbling, 4 attacking
players will start a 4 v 4 with the
defenders.
The rest of the players will
play a tennis game.

(20) ──────────── Game with a theme ──

Game with 4 color

In a field as shown in the figure, a team
is made of two colors with the condition
that the players cannot pass the ball
to a teammate with the same color.

Variation:
*Play with hands one color per team
plays with limited touches.*

(20) ──────────── Game ──

Objectives

Individual tactics:
Man marking from behind with ball passed in the air, creating space

Group tactics:

Technical training:
Running with the ball, shooting

Physical conditioning:
Aerobic power

Organization

105

 15

Warm up:
3 v 3 with two support players on each side

15

Physical:
Aerobic power

 15

Technical skill:
Running with the ball and shooting

20

Game situation:
Ball in the air

20

Game with a theme:
Marking from behind

20

Game

Warm up — 15

3 v 3 with two support players on each side
Each team has 5 players, 3 on the field and 2 on the outside.
A point is scored each time the ball is passed from one side player to the other.
The player that passes to the outside must trade place with the outside player.

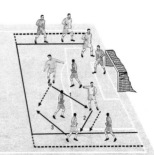

Physical — 15

Aerobic power
10:00 fartlek run
:30 of slow jog :30 of fast run

2:00 active recovery with juggling

3:00 stretching.

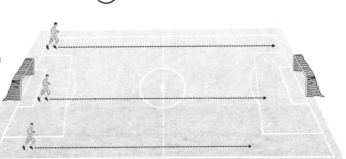

15

Running with the ball and shooting

Situation A: Running with the ball and shooting in the goal with the same color called by the coach.
Situation B: Running with the ball and shooting in goal.
Situation C: Running with the ball and shooting with right and left foot.
Shooting while turning around the cone.

20

Ball in the air

A) 1 v 1 with attacking player receiving a high ball with his shoulders to the goal. If the defender wins the ball he must dribble through the end zone to get a point. The defender cannot anticipate the opponent.

B) 1 v 1 with two goals and goalkeepers. Same activity as in A, this time if the defender wins the ball he must shoot on goal. Defender cannot anticipate.

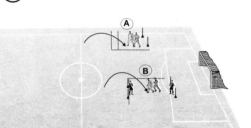

20

Marking from behind

4 v 4 or 5 v 5 is played in the central zone of the field. After three passes the team in possession of the ball must make a long pass in the air to one of the attacking players who play 1v1 with a defender. The attacking player must dribble through one of the small goals. If the defender wins the ball he will make a long pass to his team and the pass cannot be intercepted. Change the positions of the players.

20

Objectives

Individual tactics:
Man marking from behind, ball passed in the air, change of direction

Group tactics:

Technical training:
Running with the ball, long pass in the air

Physical conditioning:
Explosive power

Organization

110

15 — Warm up:
Three stations working on the weaker foot

15 — Physical:
Explosive power

20 — Technical skill:
Two situations with change of direction

20 — Game situation:
1 v 1 marking with ball in the air

20 — Game with a theme:
4 v 4 game with 4 goals

20 — Game

Warm up   15

Three stations working on the weaker foot – 5 minutes per station

a) Slalom around cones with weaker foot. 3 circuits with cones at different distances.

b) Players juggle and after 3 juggles the player receives the ball with his weaker foot – receiving the ball with inside of foot, outside of foot, instep, sole. Change method of receiving the ball every minute.

c) In pair passing and receiving with the weaker foot – volley with instep, volley with inside of foot, half volley, chest and foot.

Physical   15

Explosive power
10:00 – countermovement jump + 15m sprint. Fartlek run :30 slow jog and :30 fast run
3:00 – stretching

10:00 • 15m sprint with the ball
5:00 specific exercise – 3 skips over hurdles, heading the ball and sprint to the cone with the color called by the coach.

(20) ─────────────────────────────────── Technical skill

Two situations with change of direction

Situation A: 180 degrees change of direction.
Set up 3 cones 5 meters apart. Start dribbling, change of direction with outside foot on the first cone, with the inside foot on the second cone, sprint to the last cone and change direction with the sole of the foot and inside foot.

Situation B: Diagonal change of directions.

(20) ─────────────────────────────────── Game situation

1 v 1 marking with ball in the air

One by one the three midfielders make a long pass in the air to the zone in front of them where the players will compete in a 1 v 1.
The objective for the attacking players is to shoot in goal or to cross for the winger's, for the defenders is to mark properly and to intercept the ball.

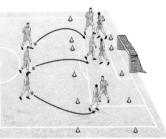

(20) ─────────────────────────────────── Game with a theme

4 v 4 game with 4 goals

Field A – 4 v 4 game with 4 goals. Limited touches to a maximum of 3 touches.

Field B – 4 v 4 game with 4 goals. Players must dribble through the goals to score and have a minimum of 3 touches.

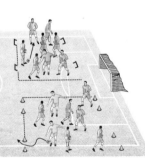

(20) ─────────────────────────────────── Game

Objectives

Individual tactics:
Man marking from behind, ball passed in the air

Group tactics:
Defensive movements of crisscross

Technical training:
Running with the ball, long pass in the air

Physical conditioning:
Quickness

Organization

110

15
Warm up:
Game of 4 v 4 with
4 colors

10
Physical:
Quickness circuit

20
Technical skill:
Dribbling and
crossing – diagonal
run and shooting

25
Game situation:
Marking with ball in
the air

20
Game with
a theme:
Defensive crisscross

20
Game

Warm up　　　　　**15**

Game of 4 v 4 with 4 colors

In a field 20 x 20 we play a 4 v 4 game
with each team made of two colors.
One color plays with limited touches
and the other color plays with
unlimited touches.

Physical　　　　　**10**

Quickness circuit

A) Low skip for 5m + lateral run around
the poles + 15m sprint with ball.
Work for 5:00.

B) Sprint with full turn around
pole + perform technical
gesture.
Work for 5:00.

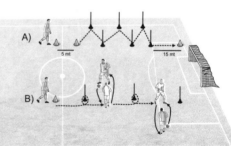

A) 5 mt　　15 mt

B)

(20) ─────────────────── Technical skill

Dribbling and crossing – Diagonal run and shooting

A group of three players starts on the wing with a ball each and must dribble around the cones and cross in the box.
At the edge of the penalty box another group of players must make a diagonal run to receive the cross and shoot in goal.

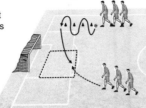
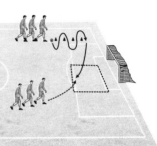

(25) ─────────────────── Game situation

Marking with ball in the air

A) Man marking on high balls from a standing position. Inside the penalty box there are three pair of players each of whom is assigned a fix marking. The winger's cross 5 teams each and the players in the box must defend and attack. After 10 crosses rotate positions.

B) 1 v 1 with attacking player with shoulders to the goal. The defender can anticipate the attacker. If the defender wins the ball he can get a point if he passes the ball to the coach on his chest. When the attacker loses possession he must prevent the defender from passing to the coach.

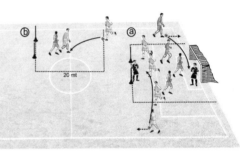

(20) ─────────────────── Game with a theme

Defensive crisscross + 2 v 2 in the penalty box

The coach sends a long ball in zone A or B where the defender starts from a higher position than his check. The center back must close down the winger and the full back must trade place with the center back as shown in the figure. The winger and the defender play in a 1 v 1 with the objective to finish with a cross in the central zone where a 2 v 2 is played.
The other players will play a 5 v 5 game with 2 touches. After 10 minutes the player switch stations.

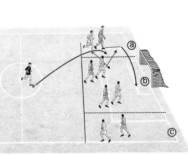

(20) ─────────────────── Game

Objectives

Individual tactics:
Feints and dribbling. Man to man marking

Group tactics:

Technical training:
Running with the ball

Physical conditioning:
Psychokinetic

Organization

110

 15
Warm up:
Game of tunnel

 15
Physical:
Game of 4 colors

 20
Technical skill:
Dribbling and tunnel

20
Game situation:
Man to man marking
on the flanks

20
Game with
a theme:
1 v 1 on the flanks

20
Game

Warm up   15

Game of tunnel

In a field of 20m x 20m we play 5 v 5.
A team score 2 points if the opponent
is beat with a nutmeg "tunnel",
otherwise the team gets only 1
point if the opponent is beat with
a dribbling. Another point is
scored every time the team
makes 5 passes.
Play for 5:00, stretch for
2:00, play for 5:00 and
stretch for 2:00.

Physical   15

Game of 4 colors

10:00 • game with 4 colors and 4 goals.
2 teams v 2 teams.
The goal is scored in the goals with
the colors of the opponents. During
the game change the alliances

2:00 • active recovery juggling

3:00 • stretching

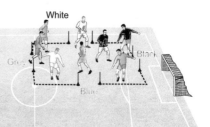

(20)

Dribbling and tunnel

Player 1 runs with the ball towards player 2 who is standing with open legs. Player 1 plays the ball through player 2' legs, collects the ball behind him, runs back to the starting position and passes the ball to player 2 who executes the same. Dribble only with right foot, only with left foot, inside and outside foot, inside of both feet "bell", only with the sole of the foot.

Variations:
Player 2 open and closes his legs in horizontal line or vertical line.

(20)

Man to man marking on the flanks

Player 1 dribbles on the flanks with the objective to cross or to beat the defender and get in the penalty box. Both actions must happen inside the 2 cones as shown in the figure. The defender cannot come out of the marked channel. The defender scores a point every time he prevents the crosses or stop the winger from dribbling into the penalty box.
Repeat the exercise from both sides of the field.

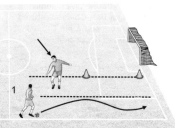

(20)

1 v 1 on the flanks

The same situation as in the previous exercise it is now progresses in a small game.
If the winger is successful with the cross a 1 v 1 is played in the penalty box, alternatively if he succeeds with the dribbling he will play a 2 v 1 in the box.
Repeat the game from both sides of the field.

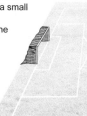
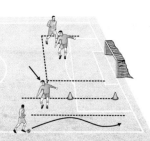

(20)

Objectives

Individual tactics:
Marking a player standing on the opposite site in relation to the ball

Group tactics:

Technical training:
Running with the ball and penetrating pass

Physical conditioning:
Psychokinetic

Organization

(110)

(15)
Warm up:
Rugby game 5 v 5

(15)
Physical:
Ball possession

(20)
Technical skill:
Running with the ball and penetrating passes

(20)
Game situation:
Passing to a teammate marked by a defender

(20)
Game with a theme:
5 v 5 game rugby style

(20)
Game

Warm up — (15)

Rugby game 5 v 5
In a field 30m x 20m we play a game rugby style, with a rugby ball. Goals are scored running through the end zone. The ball can only be passed backwards.

Physical — (15)

Ball possession
10:00 – ball possession in the central zone. At the coach's command the midfielders pass the ball to their forwards who play in a 3 v 2 and must score in the goals with the color called by the coach. Game played with no interruptions.

2:00 – recovery while juggling

3:00 • stretching.

Green

Red

Blue

Green

Red

Blue

(20)

Running with the ball and penetrating passes

On the flanks the player with the ball
dribbles around the cones and then
executes a penetrating pass towards
the goal for the winger who makes a
diagonal run and shoots towards
the far post.
In the central zone, the
penetrating pass will be for
a central striker who must
make a movement of
checking out before
receiving the ball.

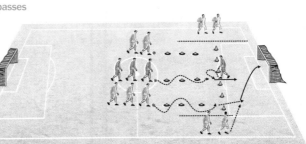

(20)

Passing to a teammate marked by a defender

The defender marks the attacker who
makes either deep runs or checks out and
checks in to receive the ball from the
midfielder. The attacking play must
score by dribbling in the end zone.
After 10:00 switch the field.

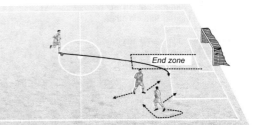

(20)

5 v 5 game rugby style

In a field 30m x 20m we play a game
rugby style, Goals are scored running
through the end zone. The ball can
only be passed backwards.
Encourage many 1 v 1 duel.

(20)

Objectives

Individual tactics:
Marking a player running from the opposite site in relation to the ball

Group tactics:
Support of the Back 4 with ball on the ground

Technical training:
Running with the ball, passing

Physical conditioning:
Psychokinetic and quickness

Organization

105

15
Warm up:
Passing and dribble in traffic

10
Physical:
Quickness circuit and technique under condition of quickness

20
Technical skill:
Marking a player coming from the opposite

20
Game situation:
Back 4 providing support with ball on the ground

20
Game with a theme:
4 v 4 with 4 goals

20
Game

Warm up   15

Passing and dribble in traffic
In a field 20m x 20m the players are in pairs and pass the ball to each other under various conditions:
a) Passing after running with the ball
b) Dribbling, feint and pass
c) Dribbling & long pass with instep
d) Hand-foot (inside)
e) Hand-foot (instep)
f) Heading the ball.

Physical   10

Quickness circuit and technique under condition of quickness
7:00 – Quick slalom around the poles with ball in the hands.
The coach calls a sequence of colors and the player will dribble through the goals following the same sequence called by the coach.

3:00 – in pairs the player perform technique under condition of quickness.

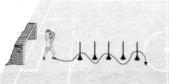

(20) ────────────────────────── Technical skill

Marking a player coming from the opposite side of the ball

A) Player 1 dribbles on the side of the penalty box and passes the ball on the ground to player 2 who runs from the opposite side. The defender must mark player 2 with the ball coming from the opposite direction

B) Same exercise as in A but the ball is crossed this time after player 1 passes the ball to himself.

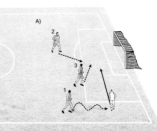

(20) ────────────────────────── Game situation

Back 4 providing support with ball on the ground

1) In the first 10:00 the defenders practice the movements of support without the ball. The players will make the movements based on where the coach will pass the ball. With the ball on the flank the coach will call whether to provide support with 2 or 3 lines. The midfielders and forwards will work on their weak foot with dribbling exercises.

2) In the next 10:00 we play a 4 v 4 focusing on the movements we worked on earlier.

(20) ────────────────────────── Game with a theme

4 v 4 with 4 goals

In a 20m x 20m field with 4 goals of different color we play a 4 v 4 with the objective of ball possession. When the coach calls a color the player in possession of the ball can score by dribbling through it (10 points) or by combination play (5 points).

(20) ────────────────────────── Game

Objectives

Individual tactics:
  Marking and creating space

Group tactics:

Technical training:
  Diagonal pass

Physical conditioning:
  Center-backs and forwards work on explosive power, midfielders work on aerobic power, wide players work on acceleration and deceleration

Organization

115

20
Warm up:
Passing and score in the small goals

20
Physical:
Differentiated work based on positions

15
Technical skill:
Diagonal pass

20
Game situation:
1 v 1 situation

20
Game theme:
Ball possession 2 teams v 1

20
Game

Warm up — 20

Passing and score in the small goals

In a field of 30m x 20m two teams of 5 players play a game trying to score by passing the ball through one of the small goals.
Set one goal more than the players in each team.

Physical — 20

Differentiated work based on positions

Centre-backs and forwards
A) 20m run while jumping + 20m sprint with ball. 10 repetitions x 2 series. 2:00 recovery – 5:00 heading in pairs.

Midfielders
B) 30m jogging + 40m of fast run + 30m jogging. 10 repetitions – 1:00 recovery.

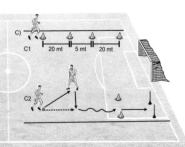

Full backs & wingers
C1) 20m sprint, stop and walk 5m, 20m sprint. 10 repetitions x 2 series. 2:00 recovery.
C2) pass the ball to the team-mate, sprint, receive the ball from the team-mate, quick dribbling and shooting in the small goal. 5:00.

15

Diagonal pass

5 players in a square pass the ball in a diagonal pattern.

Variations:
Use of left foot by changing direction of passes.
One touch passing.

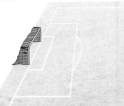
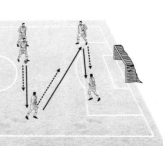

20

1 v 1 situation

A) Marking the forward while there is pressure on the ball. The forward receives the ball with his shoulders to the goal.

B) Marking the forward while there is no pressure on the ball. The forward receives the ball facing the goal and attacks the defender.
The defender must slide back to defend his zone and closing the.

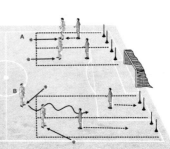

20

Ball possession 2 teams v 1

In a field of 30m x 50m two teams of 5 players each play ball possession game against one team of 5 players.
One point is scored every 10 passes.
The objective of the game is ball possession, getting open to receive the passes.
Change defending team every 6;00.

20

Objectives

Individual tactics:
Man marking with pressure and not pressure on the ball

Group tactics:

Technical training:
Passing with inside foot

Physical conditioning:
Psychokinetic specific work: center-backs and forwards work on explosive power, midfielders work on aerobic power, wide players work on acceleration and deceleration

Organization

(110)

(15)
Warm up:
Passing using a color sequence

(20)
Physical:
Differentiated work based on positions

(15)
Technical skill:
Passing

(20)
Game situation:
2 v 2 situation

(20)
Game with a theme:
2 v 2 with neutral players on the outside

(20)
Game

Warm up (15)

Passing using a color sequence
Players wear bibs of various colors and pass the ball following a sequence decided by the coach (for example, red to blue to yellow to green). Players play always with two touches. The coach changes the sequence of colors. Every two minutes 2 exercises of dynamic stretching.

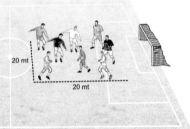

20 mt

20 mt

Variation:
Use of two balls at the same time. Play with one touch.

Physical (20)

Differentiated work based on positions

Centre-backs and forwards
A1) Squat holding the position for :10 then CMG with heading – work in pairs. 10 repetitions x 2 series. 2:00 active recovery while juggling.
A2) 20m run while jumping + 20m sprint with ball. 10 repetitions.

Midfielders
B) 30m jogging + 40m of fast run + 30m jogging. 11 repetitions :45 recovery.

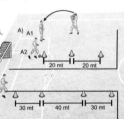

A) A1

A2

20 mt 20 mt

B)

30 mt 40 mt 30 mt

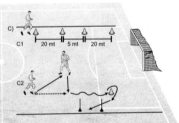

C)

C1 20 mt 5 mt 20 mt

C2

Full backs & wingers
C1) 20m sprint, stop and walk 5m, 20m sprint. 10 repetitions x 2 series. 2:00 recovery.
C2) pass the ball to the team-mate, sprint, receive the ball from the team-mate, quick dribbling with quick turn on the cone and final shooting in the small goal. 5:00.

(15) ───────────────── Technical skill

Passing
5 players in a square pass the ball in a direct pattern.

Variations:
Use of left foot by changing direction of passes. One touch passing.

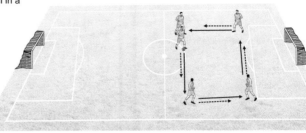

(20) ───────────────── Game situation

2 v 2 situation
One of the forward dribbles the ball backwards toward the cone creating a situation of pressure on the ball. The defenders must react to the situation in the proper way.
As the forward turns around the cone and faces the goal a situation of no pressure on the ball is created with required reaction by the defenders.
The forward can pass the ball to his team-mate creating a 2 v 2 game.

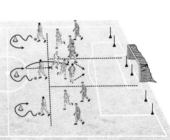

(20) ───────────────── Game with a theme

2 v 2 with neutral players on the outside
Set up more mini fields for a 2 v 2 game with 2 outside neutral players. The goal is scored by dribbling in the end zone but only after 5 passes.

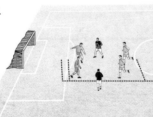
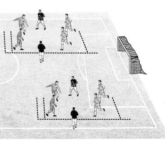

(20) ───────────────── Game

Objectives

Individual tactics:
Man marking

Group tactics:
Providing support while the ball is in the air

Technical training:
Passing and shooting

Physical conditioning:
Quickness

Organization

110

15
Warm up:
6 v 6 ball
possession game
with penetrating
passes

15
Physical:
Circuit training +
technique work

20
Technical skill:
1 v 1 situation

20
Game situation:
Providing support
when the ball is in
the air

20
Game with
a theme:
Zonal defending

20
Game

Warm up   **15**

6 v 6 ball possession game with penetrating passes

Set up a field with three zones as
shown in the figure.
A) 6 v 6 is played in the middle
zone. After 5 passes the team in
possession can pass the ball in
the attacking zone
B) for a player to run into it and
pass the ball to his captain
who is standing 5m away in
zone
C). If the final pass is
successful the team
scoring the goal remains
in possession of the
ball.
Play 2 halves of
5:00 each with 5:00
dynamic stretching
in between.

Physical   **15**

Circuit training + technique work in pairs under a condition of quickness

10:00 circuit training:
5m high skipping + 5m of low
skipping + 10m of slalom between
tight poles.
At the end of the poles sprint
to a color called by the coach
corresponding with a specific
technical gesture (e.g.
red – dribbling the ball; blue
– shooting; green – change
of direction .
5:00 of technique in
pairs under condition of
quickness e.g. perform
quick skips before
passing the ball.

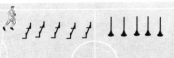

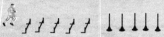

Red
Blue
Green

20

Technical skill

1 v 1 situation

A) pass the ball to the team-mate and start the 1v1. Use different ways to pass the ball (i.e. on the ground, in the air).
B) 1 v 1 in a field with two goals with goalkeepers. The ball is played by the two players on the outside. The attacking player must free himself from the defender and ask for the ball from one of the outside players. He can score in the goal indicated by the direction as shown in the figure. If the defender wins the ball he becomes the attacker. Change exercise after 10:00

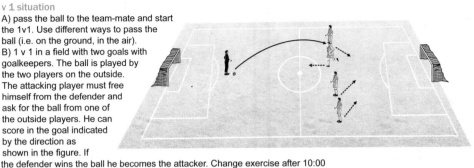

20

Game situation

Providing support when the ball is in the air

Only the defenders are involved in this exercise. The back 4 train on providing support to the player attacking the ball in the air, as shown in the figure.
After 10:00 introduce two forwards, one of them tries to flick the ball for the other forward running into the space. The other players play a 5 v 5 game with four goals, with two touches and with the objective of keeping possession of the ball.

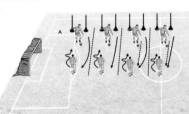

20

Game with a theme

Zonal defending

The blue team after 5 passes can pass the ball in Zone A (2 v 2) or Zone B (1 v 1). In zone A one of the forward must receive the ball first.
In zone B the wing scores a point if he can cross the ball or if he dribbles through the cones and gets into zone A where the objective is to score in goal.
If the white team wins the ball, after 5 passes they can score by dribbling through one of the three goals.
Change half after 10:00.

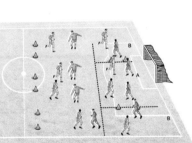

20

Game

Objectives

Individual tactics:
Creating space

Group tactics:
Quick counter-attacks

Technical training:
Penetrating passes

Physical conditioning:
Center-backs and forwards work on explosive power, midfielders work on aerobic power, wide players work on acceleration and deceleration

Organization

110

20
Warm up:
4 v 4 v 4 ball possession game

15
Physical:
Differentiated work based on positions

15
Technical skill:
"Give and go" pass for quick counter-attacks

20
Game situation:
2 v 1 tournament

20
Game with a theme:
3 v 3 with support players on the outside

20
Game

Warm up   20

4 v 4 v 4 ball possession game
Set up a field long and narrow. The objective is to make a flighted pass to the team in the opposite side of the field after 6 passes.

The team that cannot accomplish the objective becomes the defending team applying pressure and intercepting the ball.

Physical   15

Differentiated work based on positions
Centre-backs and forwards
A1) 10 squats x 4 series – :30 recovery
A2) Jump three overs and on the fourth over head the ball launched by the coach 10 repetitions for 2 series.

B) Midfielders
7:00 running around the field alternating :45 of jogging to :15 of fast run. 2 series with 2:00 recovery.

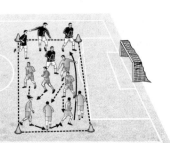

Full backs and wingers
C1) 15m sprint, stop in the square and sprint towards the cone called by coach: 10 repetitions for 2 series. 2:00 recovery.
C2) dribbling over 60m with coach calling "stop" and "go". 5 repetitions for 2 series. 2:00 recovery.

(15)

"Give and go" pass for quick counter-attacks

In 15m x 10m field two players act as walls on each side of the field. Player A1 starts by passing to the outside player to receive in A2 and pass to B who performs the same task. Player A applies passive pressure.

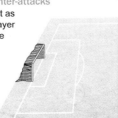
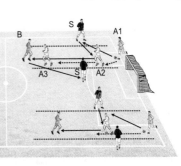

(20)

2 v 1 tournament: Creating numerical advantage

Set more fields 20m x 10m with two zones. In each zone we play a 1 v 1. Once the ball is passed to the teammate in the other zone the player can join in the attack to form a 2 v 1.

Variation:
2 v 2 game.

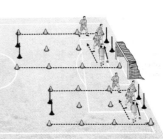

(20)

3 v 3 with support players on the outside

Set up a 30m x 30m field. We play a 3 v 3 game and each team has two outside players.

A goal is scored every time the ball is passed from one end to the other end of the field using the outside players.
A player who passes the ball to the outside player takes his position.

(20)

Objectives

Individual tactics:
Creating space

Group tactics:

Technical training:
Penetrating passes

Physical conditioning:
Center-backs and forwards work on explosive power, midfielders work on aerobic power, wide players work on acceleration and deceleration.

Organization

110

(20)
Warm up:
3 v 3 v 3 speed of thought

(15)
Physical:
Differentiated work based on positions

(15)
Technical skill:
Penetrating pass, wall pass and shooting

(20)
Game situation:
2 v 2 with deep target player

(20)
Game with a theme:
3 v 3 with 2 target players in the end zone

(20)
Game

Warm up   (20)

3 v 3 v 3 speed of thought
Set up a field of 20m x 20m
Two teams of different color play against one team creating a 6 v 3.
The objective is to make 10 continuous passes.
The team loosing possession becomes the defending team.

Physical   (15)

Differentiated work based on positions
Centre-backs and forwards
A1) 10 squats x 5 series :30 recovery
A2) Jump and head the ball three times. 10 repetitions for 2 series.

Midfielders
7:00 running around the field alternating :40 of jogging to :20 of fast run. 2 series with 2:00 recovery

Full backs and wingers
C) Sprint to the cone, stop and walk to the
teammate to receive and pass the ball, continue doing the same to the other cones. 10 repetitions for 2 series with 2:00 recovery.

15 ───────────────────────────────── Technical skill

Penetrating pass, wall pass and shooting

Two groups of 5 players compete in a
game of passing and shooting. Player
A passes to B who performs a wall
pass for the incoming A who
shoot in goal.

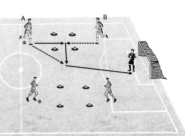

Variations:
Wall pass with only one touch. Flighted pass in the square receiving ball in the air.

20 ───────────────────────────────── Game situation

2 v 2 with deep target player

In various mini field we play 2 v 2 games
with one of the attacking players playing
as a target player trying to wall pass
the ball.
If the defenders intercept the ball
they score by dribbling over the
end zone.

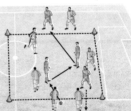

20 ───────────────────────────────── Game with a theme

3 v 3 with 2 target players in the end zone

Set up a 30m x 30m field.
We play 3 v 3 game and each team has
a target player in both end zones.
A goal is scored every time the ball
is passed from one end to the
other end of the field using the
target players.

20 ───────────────────────────────── Game

Objectives

Individual tactics:
Creating space

Group tactics:
Providing defensive support while the ball is in the air - playing out from the back

Technical training:
Passing and shooting

Physical conditioning:
Quickness

Organization

(110)

 (15)
Warm up:
5 v 2 using two touches

 (15)
Physical:
Differentiated work based on positions

 (20)
Technical skill:
2 v 1 exercises

(20)
Game situation:
Defensive support with ball in the air + 2 midfielders

(20)
Game with a theme:
6 v 6 v 6: shot on goal within 5 passes

(20)
Game

Warm up   (15)

5 v 2 using two touches

Fields of 10m x 10m we play a 5 v 2 with two touches. Three series of 3:00 each. In between series 3:00 of general mobility and stretching

Physical   (15)

Differentiated work based on positions

A) 5:00 of technique in pairs under condition of quickness: low skipping (forward, backward, forward) then execute the technical gesture. Each player works for :30. Players execute the following:
 • Volley inside foot left and right
 • Volley instep left and right
 • On the ground inside foot left and right
 • Half volley left and right foot
 • Heading.
B) 3:00 of low skipping over 5 obstacles followed by 5m of light jogging.

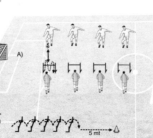

5 mt

(20)

2 v 1 exercises

A) Central defenders and forwards work on 2 v 1 in the central zone of the penalty area. Forwards have 4 seconds to shoot.
B) Remainder of the players work on situations of 2 v 1 with overlapping. The attacking players can score a point if they are able to cross the ball in zone 1 or zone 2 as it is called by the goalkeeper.

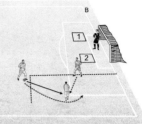

Variations for A:
1. Forwards start facing the goal 2. One forward starts with his back to the goal 3. One forward enters the field from the outside making a diagonal run.

(20)

Defensive support with ball in the air + 2 midfielders

We work with the proper movement of the 4 defenders in providing support to the defender attacking the ball in the air. One defender attacks the ball and the other 3 defenders drop off to provide support. Two midfielders are introduced to work on play build up from the back.
1. Build up from the back 4 using the central midfielders
2. Build up with the central defender breaking through
3. Moving the ball across the back four using the central midfielders.

(20)

6 v 6 v 6: shot on goal within 5 passes

6 v 6 game as shown in the figure. If a goal is scored within 5 passes the team gets a point. Every time a team takes a shot on goal that team get out of the field and the team waiting on the sideline enters the field. Wins the team that score more goals.
Play 2 halves of 8:00 each.
A quick change is required by the teams.

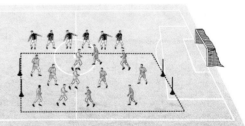

(20)

Objectives

Individual tactics:
  Creating space

Group tactics:
  Ball possession

Technical training:
  Penetrating passes

Physical conditioning:
  Center-backs and forwards work on explosive power, midfielders work on aerobic power, wide players work on acceleration and deceleration

Organization

110

15
Warm up:
5 v 3

15
Physical:
Differentiated work based on positions

20
Technical skill:
Penetrating passes

20
Game situation:
3 v 2

20
Game with a theme:
4 v 4 tournament

20
Game

Warm up — 15

5 v 3

Three defenders play against five attackers. The attacking team scores a point if they...
  • Make ten consecutive passes
  • Score in the goal

The defenders take away 2 points to the attacking team if they win the ball and dribble through the end zone.

Variations:
1. Two touches for the attacking team
2. The defender that wins the ball must dribble trough the end zone with no support from his teammates.

Physical — 15

Differentiated work based on positions

Centre-backs & forwards

A1) Countermovement jump + 10m sprint & stop. 3 series of 10 repetitions with :15 recovery. 1:00 recovery between series

A2) Countermovement jump with sprint and change of direction (deceleration at the dark cone, light jog and change of direction at the light cone and sprint again to the next cone) Circuit with 4 sprints. 2 series of 10 repetitions. :20 recovery between repetitions. 1:30 recovery between series.

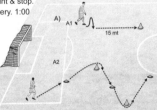
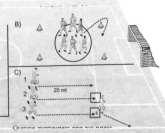

Midfielders

B) inside the circle players with various color bibs pass the ball around following a specific color sequence. Players must be moving at all times. Every :30 players must sprint outside the circle to a cone and return inside the circle. 2 series of 6:00 with a 3:00 recovery. Full backs and wingers – C1) 1.20m sprint leaning forward first   C2) 20m sprint leaning forward first and stop in zone A C3) 20m sprint leaning forward first with stop in zone A and change of directions. 10 repetitions for each exercise.

20 — Technical skill

Penetrating passes

6 players dribble the ball in the centre circle. Each player has a teammate that dribbles the ball in one of the lateral zones as shown in the figure. At discretion of one of the lateral players, he leaves the ball and makes a run towards one of the boxes at the edge of the penalty area.
This player must receive a pass from his teammate in the centre circle who should have noted the diagonal run.

20 — Game situation

3 v 2

A) 3 v 2 with the forwards facing he goal. The attacking team has 5 seconds to score a goal. If the defenders win the ball they score by dribbling trough the end zone.
B) 3 v 2 with on forward receiving the ball with his back to the goal. Defenders can score as in A)
C) 3 v 2 with one defender playing as a sweeper behind the other defender. The attacking player in possession of the ball must ensure to pass the ball to the open teammate.

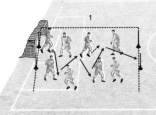
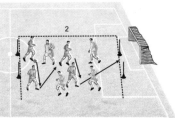

20 — Game with a theme

4 v 4 tournament

4 v 4 tournament with conditions
1) Score a goal within 4 passes
2) Score a goal after 4 passes.

20 — Game

Objectives

Individual tactics:
 Creating space

Group tactics:
 Quick play

Technical training:
 Penetrating passes

Physical conditioning:
 Center-backs and forwards work on explosive
 power, midfielders work on aerobic power, wide
 players work on acceleration and deceleration

Organization

115

20
Warm up:
6 v 6 handball
game

15
Physical:
Differentiated work
based on positions

20
Technical skill:
Diagonal run
and volley

20
Game situation
3 v 3 v 3 in the
penalty area

20
Game with
a theme:
8 v 4

20
Game

Warm up   20

6 v 6 handball game

In a field as shown in the figure we
play a 6 v 6 handball game.

The goal must be scored with a
volley. If the player in possession
of the ball is touched by the
opponent there is ball turnover.
Quick play is encouraged.

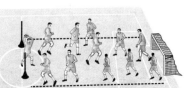

Physical   15

Differentiated work based on positions

Centre-backs & forwards

A1) In pair squat keeping the position for :10,
jump and heading the ball. 2 series of 20
repetitions each. Recover of 1:00. A2)
Countermovement jump + 10m sprint and
stop. 3 series of 10 repetitions with :15
recovery. 1:00 recovery between series

Midfielders

B) juggling the ball inside the
square. Every :30 the players
run around the perimeter of
the area at maximum speed
for :30. 2 series of 6:00
with 2:00 recovery.

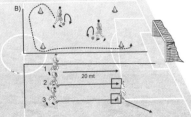

Full-backs & wingers

C1) 1.20m sprint leaning forward first, with the ball   C2) 20m sprint leaning forward first with the ball and stop in zone A
C3) 20m sprint leaning forward first with the ball, stop in zone A and change of directions. 10 repetitions for each exercise.

20 ——————————————— Technical skill

Diagonal run and volley

Players dribble around the cones and lobby the ball in the opposite square where the teammate makes a diagonal run to receive the ball and shoot with a volley.

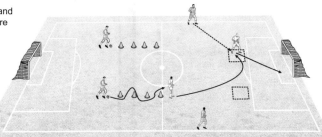

Variations:
a. Heading the ball
b. Control the ball and shoot
c. Shoot with a Half volley.

20 ——————————————— Game situation

3 v 3 v 3 in the penalty area

Three teams of three players each play a game with the condition to shoot in goal after three passes.

20 ——————————————— Game with a theme

8 v 4

In a field long and narrow we play 4 v 4 with 4 support players on the outside who play with the team in possession of the ball.

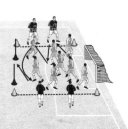

Variations:
Limited touches for every player
The support players can pass the ball to each other.

20 ——————————————— Game

Objectives

Individual tactics:
  Creating space

Group tactics:
  Providing defensive support while the ball is in the air and build up from the back

Technical training:
  Penetrating passes

Physical conditioning:
  Speed of thought and quickness

Organization

110

15
Warm up:
3 v 1

15
Physical:
Differentiated work based on positions

20
Technical skill:
2 v 1 with diagonal run and passing

20
Game situation:
Defensive support

20
Game with a theme:
4 v 4 + 4 support players

20
Game

Warm up

15

3 v 1

3 v 1 keep away. The defending players must intercept as many balls as possible in 1:00.
The other players must always create space by moving along the side of the field.

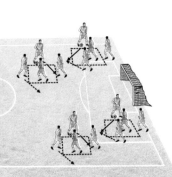

Physical

15

Differentiated work based on positions

A) Three players work on technique in pairs under condition of quickness: low skipping then execute the technical gesture. Each player works for :30. Players execute the following:
  • Volley inside foot left and right
  • Volley instep left and right
  • On the ground inside foot left and right
  • Half volley left and right foot
  • Heading.

B) 2:00 of low skipping over 5 obstacles followed by 5m of light jogging.

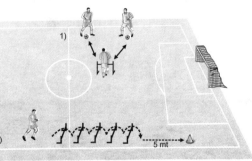

5 mt

(20) ────────────────────────── Technical skill

2 v 1 with diagonal run and passing

Organize the field as shown in the figure. Midfielder dribbles around the cones and lobbies the ball in one of the square chosen by the run of the forward player who is checked by the defender. The forward player can do one of the following:
A) turn and shoot
B) Play 2 v 1 with the support of wing player who has made a diagonal run inside the field.

(20) ────────────────────────── Game situation

Defensive support with ball in the air + 2 midfielders+ 2 forwards

We work with the proper movement of the 4 defenders in providing support to the defender attacking the ball in the air. One defender attacks the ball and the other 3 defenders drop off to provide support. Two forwards are introduced; one tries to flick the ball with his head for the other forward who has run into the space created by the other forward. If the defenders recover the ball they must dribble through the end zone with the full back players with the support of the central

midfielders. The forward players must prevent this from happening. Rules of the game: always let the forward flick the ball, the defense must play from the back using ball possession.

(20) ────────────────────────── Game with a theme

4 v 4 + 4 support players in the attacking zone

In a field as shown in the figure we play 4 v 4 with 2 goalkeepers. Each team has 4 support players in the attacking zones, 2 on the side and 2 in the end zones.

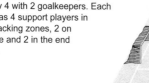
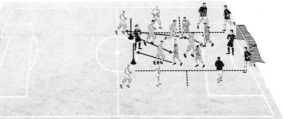

Variations:
 • *Limited touches for support players.*
 • *Play with hands.*

(20) ────────────────────────── Game

Objectives

Individual tactics:
Creating space

Group tactics:

Technical training:
Penetrating passes

Physical conditioning:
Psychokinetic

Organization

110

15	10	20
Warm up:	Physical:	Technical skill:
4 v 4 diamond	4 v 4	Penetrating pass with shielding of the ball

20	25	20
Game situation:	Game with a theme:	Game
3 v 2 with penetrating passes	Attacking the space	

Warm up — 15

4 v 4 diamond

4 v 4 game keeping a diamond shape with the objective to score goals by dribbling in the end zones focusing on playing off the target player.

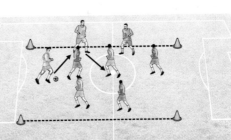

Physical — 10

4 v 4

4 players with different color bibs in each field. After 3 passes between players in field A the ball is passed in the air to field B to the player with the color of the player in A, who has not touched the ball. 5:00 with one ball and 5:00 with 2 balls.

20 ────────────────────────────────── Technical skill

Penetrating pass with shielding of the ball

Players are divided in two groups and play a 2 v 1 focusing on penetrating passes. The forward creates space for himself and makes the movement to receive the ball from the teammate who can enter the field only after the pass.

20 ────────────────────────────────── Game situation

3 v 2 with penetrating passes

Set two fields where a 3 v 2 is played. One of the forward acts as target player. He must receive the ball from the other teammate who then make runs to create space for themselves. The goal is scored by dribbling through the end zones.
If the defenders intercept the ball and score by dribbling in the other end zone they get two points.

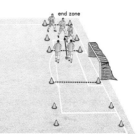

25 ────────────────────────────────── Game with a theme

Attacking the space

Set a field as shown in the figure. 5 v 5 with 2 support players per team on the sidelines.
The goal is scored when a penetrating pass is made between the cones in the end zone and a teammate makes a run to receive the pass.

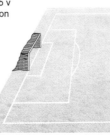
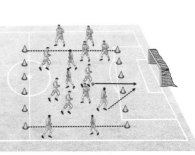

20 ────────────────────────────────── Game

Objectives

Individual tactics:
Creating space

Group tactics:

Technical training:
Passing and ball control

Physical conditioning:
Coordination skills of adaptation and
transformation, differentiation and orientation

Organization

115

15
Warm up:
Juggling in pair

20
Physical:
Exercise with
4 teams

20
Technical skill:
Passing among
three players

20
Game situation:
2 v 2 + 2 support
Players + 2 v 2

20
Game with
a theme:
8 v 8 + 2
goalkeepers

20
Game

Warm up   **15**

Juggling in pair
In pairs the players perform various
type of juggling...
  • Free
  • 2 touches
  • Head, chest and foot only.
  • With two balls.
  • With mandatory use of the
thigh to receive the ball.
  • With balls of various
dimensions.

Physical   **20**

Exercise with 4 teams
Every team has available 15 passes
in the air to the team with the color
called by the coach.
The team earns a point if the ball
ends inside the field.
Change the type of pass and
the type of commands, audio
and visual.

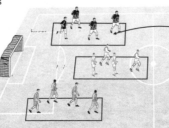
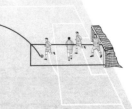

(20) ───── Technical skill ─────

Passing among three players

Quick passes of give and go and movement. Each player must move in the center.

Variations:
• *Passes on the ground*
• *Passes in the air (players are wide)*
• *One touch passes.*

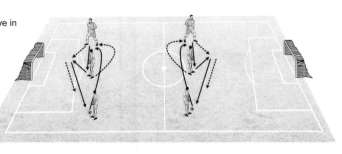

(20) ───── Game situation ─────

2 v 2 + 2 support players + 2 v 2

A) Set a field 10m x 10m for a 2 v 2 with 2 support players on the outside for the team in possession of the ball.
A point is scored every time the ball is passed from one support player to the other using the inside players.

B) a double 2 v 2 is played, with team A playing against team B from right to left, and team C playing against team D from top to bottom. Use two balls.

(20) ───── Game with a theme ─────

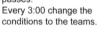

8 v 8 + 2 goalkeepers

One team must play ball possession and try to shoot only after a predetermined number of passes. The other team instead must score with a maximum of 4 passes.
Every 3:00 change the conditions to the teams.

(20) ───── Game ─────

Objectives

Individual tactics:
Creating space

Group tactics:
Sliding backwards of the defensive unit

Technical training:
Passing

Physical conditioning:
Quickness

Organization

110

15 Warm up:
Ball possession

15 Physical:
Differentiated work based on positions

20 Game situation:
4 v 2 + 2 support players + 2 v 2 + 1

20 Tactical situation:
Delay & contraction of the defensive unit

20 Game with a theme:
6 v 4 on two fields

20 Game

Warm up　**15**

Ball possession
5 v 5 v 5 ball possession. Two teams play against one. The only objective of the defending team is to win the ball. Every team will defend for 5:00. Team that allows the least passes wins.

25 mt
25 mt

Physical　**15**

Differentiated work based on positions
1) The player passes the ball with his hand to his teammate and performs a low skipping. 10 passes and sprint right, 10 passes and sprint left, 10 passes and sprint back. 4 series of each.

2) Low skips and sprint towards the color called by the coach. 3:00.

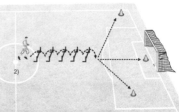

(20) ————————————————— Game situation —

4 v 2 + 2 support players + 2 v 2 + 1

A) Set a field 10m x 10m 2 v 2 inside the field and each team has 2 support players on the outside. We create a 4 v 2. A point is scored when the ball is passed from one support player to the other passing through the two midfielders. Play for 10:00.
B) Set up a field 10m x 10m 2 v 2 with a neutral player who plays with the team in possession of the ball. Score the goal by dribbling through the end zone but both players must touch the ball.

(20) ————————————————— Tactical situation —

Delay and compressing movement of the defensive unit

Set the back 4 line with 4 midfielders at center half. The back 4 build up the play using the width but when the reach the center half they voluntarily give the ball to the midfielders. The player receiving the ball will attack the space in front of him and the back 4 will run backwards delaying the action of the midfielder, until they reach the edge of the penalty area. At this point they will close the space using the proper zonal defending movements.

(20) ————————————————— Game with a theme —

6 v 4 on two fields

6 v 4 as shown in the figure. If the team scores a goal, we start the game again from the opposite field. If the defenders win the ball they must move the ball until the center half with the only pressure created by the 2 forwards (the 4 midfielders remain passive). If they succeed in reaching the half they pass the ball to their midfielders who start the attacking play in the other field. If the two forwards win the ball the 4 midfielders become active and we start a 6 v 4 in the first field. An unsuccessful shot on goal results in the turn over to the other team.

(20) ————————————————— Game —

Objectives

Individual tactics:
Creating space, crossing

Group tactics:

Technical training:
Long passes

Physical conditioning:
Anaerobic endurance

Organization

(110)

(15)
Warm up:
4 v 4 with long pass

(15)
Physical:
Running with the ball

(20)
Technical skill:
Long passes in the square

(20)
Game situation:
Long pass and 2 v 1

(20)
Game with a theme:
8 v 8 in the central zone

(20)
Game

Warm up   (15)

4 v 4 with long pass

4 v 4 ball possession in a 20m x 20m field. The player in possession of the ball must send a long ball one of the players in the outside fields with a color called by the coach. The point is scored if the player in the outside the box receives the ball correctly.

Physical   (15)

Running with the ball

As shown in the figure, player A and B run with the ball to the cones in front of them, where they leave the ball and sprint towards cone 3 (player A) and cone 2 (player B). Here they start dribbling with the ball again to the next cone and sprint back, for a total of 10 diagonal sprints.

Perform 4 series.
Recover of 1:30.

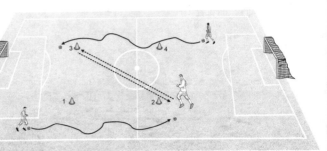

(20) ———————————————— Technical skill

Long passes in the square
Player A makes a long pass in the air
to player B.
If B does not control the ball in the
square player A gets a point.

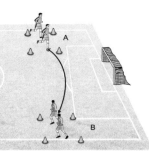

(20) ———————————————— Game situation

Long pass and 2 v 1
Player A makes a long pass to his
teammate who can receive the ball
and shield for 5" or make a wall
pass for Player A. The player
receiving the ball must perform
a long/short movement.
The attacking players must
score in the end zone, if
the defender intercepts
the ball he can score
in one of the two
goals.

(20) ———————————————— Game with a theme

8 v 8 in the central zone
Players are placed as shown in the figure.
After 5 passes the team in possession
of the ball can make a long pass in
one of the two zones A where one
player can run into with no players
from the opposing team. The
player receiving the ball in zone
A must cross the ball in zone
B where only two attacking
players and one defender
are allowed in.

(20) ———————————————— Game

Objectives

Individual tactics:
Creating space

Group tactics:

Technical training:
Long passes

Physical conditioning:
Speed endurance

Organization

120

15
Warm up:
4 v 4 + 4 support
players

20
Physical:
Athletic exercise

20
Technical skill:
Long passes to a
running teammate

20
Game situation:
Long pass
& 2 v 1

25
Game with
a theme:
6 v 6 and long
pass for a running
teammate

20
Game

Warm up **15**

4 v 4 + 4 support players with pass to goalkeeper
4 v 4 game with 4 support players on
the outside. The objective is to create
space to receive the ball and make a
long pass to the goalkeeper who is
standing 15m away from the field.

Physical **20**

Athletic exercise
10 repetitions of sprints between
distances of 5m
:30 recovery between repetitions.
2 series with recovery of 1:00.

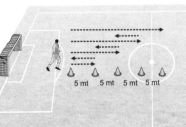

5 mt   5 mt   5 mt   5 mt

(20) ─────────────────────────────────── Technical skill

Long passes to a running teammate

Set many 20m x 20m mini fields. 2 players in each field work on long passes and receiving the ball. Player A dribbles the ball to the next cone makes a long diagonal pass to player B who receives and dribble to the next cone. He performs the same pass.

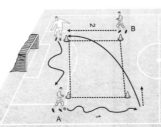
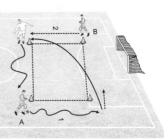

(20) ─────────────────────────────────── Game situation

Long pass and 2 v 1

One player makes a long pass in the box. The player receiving the ball plays a wall pass to one of the two supporting lateral players and then play in a 2 v 1 trying to score in goal. If the defender intercepts the ball he can score in one of the two mini goals.

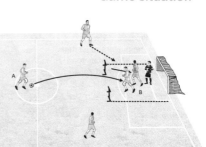

(25) ─────────────────────────────────── Game with a theme

6 v 6 and long pass for a running teammate

6 v 6 game. After 5 passes the team in possession of the ball can make a long pass in the lateral channel for a player who makes a run from behind without being challenged by any defenders as shown in the figure).
If the pass is successful and the wing player receives correctly he can cross the ball. Two points for a goal from a cross.

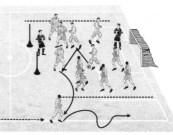

(20) ─────────────────────────────────── Game

Objectives

Individual tactics:
Creating space, diagonal runs

Group tactics:
Attacking and defending with build up from the back

Technical training:
Passing and shooting

Physical conditioning:
Quickness

Organization

110

(15)
Warm up:
2 v 2 + neutral player

(15)
Physical:
Differentiated work based on positions

(20)
Game situation:
Dynamic 2 v 2

(20)
Tactical situation :
Shooting

(20)
Game with a theme:
From 6 v 4 to 4 v 6

(20)
Game

Warm up   (15)

2 v 2 + neutral player

Set many mini fields. 2 v 2 with a neutral player. The players must make 2 or 3 passes before they can score by dribbling through the end zones.

Variations:
• *Unlimited touches*
• *Two touches.*

Physical   (15)

Differentiated work based on positions

1st exercise: 15m of lateral movements and quick turn around the pole. 10 repetitions

2nd exercise: 15m of lateral movements and quick turn around the pole with summersault and spring at the end of the circuit. 5 repetitions

3rd exercise: 5m of backward run and sprint forward.

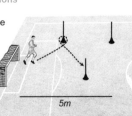

5m

(20) — Game situation

Dynamic 2 v 2

Players A and B are turned away from the goal. At the audio signal they must turn and the first player on the ball can pass to any of player C or D who becomes his teammate. The other player becomes a defender with the player who arrived second on the ball.

Variations:
- Visual signal
- Passing the ball with feet
- Passing the ball with head.

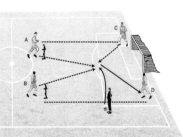

(20) — Tactical situation

Shooting

A) Wing player makes a diagonal run to receive the ball in the center of the field and shot in goal with the pressure from the defender who starts from an outside position.

B) various shooting...
- Juggling & shooting
- Dribbling & shooting
- Dribbling feint & shooting

Variation:
(For exercise A) Active defender - Long pass in the air.

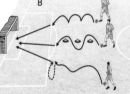
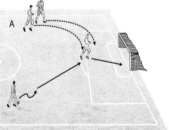

(20) — Game with a theme

From 6 v 4 to 4 v 6

4 defenders play against 4 midfielders and 2 forwards. The defenders work on zonal defending and when they intercept the ball they can score by playing the ball in the opposite site using the support of the two center midfielders with pressure from the two forwards and the two wing players.

(20) — Game

Objectives

Individual tactics:
Creating space and support

Group tactics:
Movements of central defenders and central midfielders

Technical training:
Passing and receiving

Physical conditioning:
Anaerobic endurance

Organization

110

10
Warm up:
4 v 4
ball to support
player

20
Physical:
3:00 of run with
recovery

20
Technical skill:
Creating angles of
support

20
Game situation:
2 v 1 creating
angle of support

20
Game with
a theme:
5 v 5 with pass to
support player

20
Game

Warm up

10

4 v 4 ball to support player
Set a 20m² field where we play a 4 v 4 game + 1 player in a supporting role. One point is scored every 10 passes and every time the ball is passed back to the supporting player and he passes it back to a teammate.

Physical

20

3:00 run with recovery
3:00 of continuous run at 90% (maximum intensity). 3 series with 3:00 recovery.

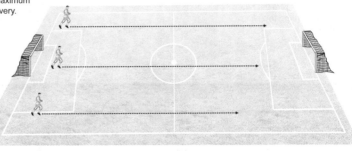

20

Technical skill

Creating angles of support and proper movements of central defenders and central midfielders

In each mini field there are three mini goals with three players standing behind the goals and play the ball around. The outside player plays the ball to the closest player inside the field and applies pressure. The player receiving the ball must pass to his teammate in a support position and this must score in the mini-goal that has been left unattended from the shifting of the other players.

20

2 v 1 creating angle of support

Set a few mini fields where we play a situation of 2 v 1 with 2 support players.
A point is scored every time the ball is passed from one support player to the other but only after a "give and go" of the two inside players.

20

5 v 5 with pass to support player

Set a 30m² field where we play a 5 v 5 game.
A goal is scored by dribbling through the end zone.
The forward pass is not allowed but the players must use the pass to the support player to gain the open space.

20

Objectives

Individual tactics:
Creating space

Group tactics:

Technical training:
Ball control and shooting

Physical conditioning:
Speed endurance

Organization

110

20
Warm up:
Passing sequence
with three players

15
Physical:
Sprint to the color
+ wind sprints

20
Technical skill:
Shooting

15
Game situation:
5 v 3 + 1
Goalkeeper

20
Game with
a theme:
4 v 4 tournament
with a theme

20
Game

Warm up — 20

Passing sequence with three players

As shown in the figure player A passes
to B who returns the ball to A, A then
passes to player C and trade place
with player B.

Variations:
- *Unlimited touches*
- *Two touches*
- *One touch*
- *Ball on the ground with two touches*
- *Ball on the ground with one touch.*

Physical — 15

Sprint to the color + wind sprints

1st exercise: Sprint towards the color
called by the coach and light jog
back. 10 repetitions 5 series.

2nd exercise: wind sprints over
20m 10 repetitions with :15
recovery. 2 series with 1:
recovery.

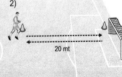

(20) ——————————————— Technical skill

Shooting

A) Shooting after a circuit of mobility and quickness. In both circuits a string is placed between two poles, one 30cm high and the other 120cm high. The players must dribble by skipping over the string or ducking under it.

B) Shooting in goal overtaking a passive defender.

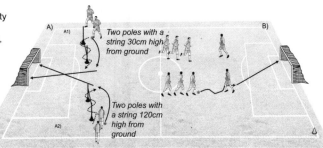

Two poles with a string 30cm high from ground

Two poles with a string 120cm high from ground

Variation for B)
Juggling • Overtaking the defender with a nutmeg • Overtaking the player with a "rainbow move"

(15) ——————————————— Game situation

5 v 3 + 1 Goalkeeper

5 v 3 with the goalkeeper playing with the team of 3 players. The team of 5 players score a point every 10 passes or every time it scores in the small goal. The team with 3 players scores a goal every time the ball is passed to the goalkeeper.
Change teams every 5:00.

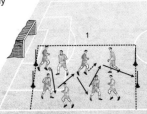
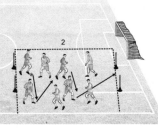

(20) ——————————————— Game with a theme

4 v 4 tournament with a theme

4 v 4 tournament with theme
1st theme: A goal can be scored only after 4 passes
2nd theme: A goal must be scored only within 4 passes.

(20) ——————————————— Game

Objectives

Individual tactics:
Creating space, countermovement

Group tactics:
Defensive elastic and simple attacking solutions

Technical training:
Long pass

Physical conditioning:
Quickness

Organization

110

15
Warm up:
4 v 4

10
Physical:
Exercise to be performed in pair

20
Game situation:
2 v 1 with countermovement and creating space

20
Tactical situation:
Group tactics

25
Game with a theme:
Countermovement
• creating space
• 2 v 2

20
Game

Warm up   **15**

4 v 4
Start with a simple 4 v 4. Every goal that is the result of a countermovement (going long and short) will be awarded 2 goals. After 8:00 switch to play with a diamond with the target player playing only as a wall and laying off the ball to the other 3 players.
Emphasize the countermovement.

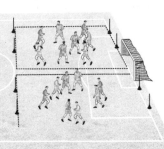

Physical   **10**

Exercise to be performed in pair
Exercise to be performed in pair. Players start with a summersault, skip over the circles and the first to arrive on the ball can shoot in goal.

After 3:00 add a pole that the player with the ball must dribble around it.

After 3:00 add some obstacles for low skip.

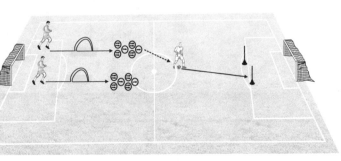

20 — Game situation

2 v 1 with countermovement and creating space

Players are divided in groups of three.
Set more mini fields as in the figure.
One of the attacking players perform
a countermovement to create
space, he receives the ball and
lay back to the teammate to
start a 2 v 1 with the objective
to shoot in goal within :05.
If the defender wins the
ball he can dribble over
the end zone.

Variations:
Target player use one touch for the wall pass, he uses his head, he controls the ball and pass back.

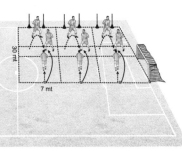

20 — Tactical situation

Group tactics

Tactical exercise for all defenders. The
defensive unit is placed as shown in
the figure. Players A-B-C-D move the
ball around and the 4 defenders
must move as a result adopting
the concepts of pressure on the
ball and no pressure on the ball
(therefore closing down the
space or dropping back).
The defenders also provide
the proper support when
the ball is on the lateral
players C-D.
The other players will perform exercises of dribbling and running with the ball.

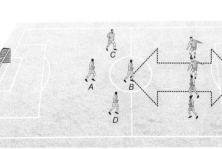

25 — Game with a theme

Countermovement – creating space – 2 v 2

Set the players as shown in the figure.
Midfielder dribble with the ball and one of
the strikers makes a countermovement
and enters into one of the square
where a defender cannot get in. The
striker lays the ball off for one of
the wingers who makes a run
inside the filed and shoot in goal.
Right after the shoot the coach
crosses the ball for a 2 v 2
between the 2 strikers and
the 2 central defenders.

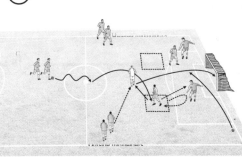

20 — Game

Objectives

Individual tactics:
  Dribbling and feints

Group tactics:

Technical training:
  Shooting with weaker foot

Physical conditioning:
  Psychokinetic – speed endurance

Organization

105

10
Warm up:
Game of the 4 colors

15
Physical:
Juggling in pairs

20
Technical skill:
Shooting

20
Game situation:
1 v 1

20
Game with a theme:
5 v 5 quick shooting

20
Game

Warm up　　　　**10**

Game of the 4 colors

Divide the players in 2 groups of 8-10 players. Divide each group in 4 colors. We play with hands with the following rules and variations:
a) Player passes the ball and calls the color he passes to
b) Player passes the ball and calls a color different then the color he passes the ball to.
c) Player passes the ball and calls the color that the player receiving the ball must pass to.
d) Player passes the ball and calls a color that the player receiving the ball cannot pass to.

Physical　　　　**15**

Juggling in pairs

The player passes the ball to the teammate and runs around the cone. The other player juggles the ball as he waits.

Series of 3:00 with 1:30 recovery.

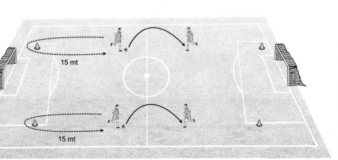

15 mt

15 mt

20 — Technical skill

Shooting

Place 2 lines of 5 poles or cones (50cm apart) 10m outside the penalty area. Player A passes the ball to himself, he starts running around the cones and shoot in goal. Shooting with right foot from the left side and left foot from the right side.

Variations:
- *Running backwards*
- *– Run from the central position*
- *– Run from the lateral sides*
- *– Shoot a bouncing ball after the player self passes the ball from a throw in.*

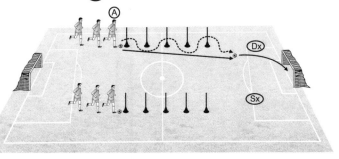

20 — Game situation

1 v 1

The players skip over the obstacles and try to win a ball kicked in the field by the coach. The player who wins the ball shoots in goal. If the goalkeeper saves the shot he immediately passes to his teammate who now attacks the other goal. The coach plays 3 balls for each pair.

20 — Game with a theme

5 v 5 quick shooting

In a 10m x 15m field two teams of 5 players each try to continuously shoot in goal by quick combination plays. Goalkeeper can distribute high ball in order to allow acrobatic shots. The small field is to promote quick shooting.
The remainder of the squad performs exercises of dribbling with left foot.

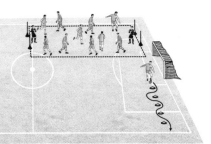

20 — Game

Objectives

Individual tactics:
Dribbling and feints

Group tactics:

Technical training:
Shooting and heading

Physical conditioning:
Speed endurance

Organization

115

15
Warm up:
2 games of 4 v 4

20
Physical:
Athletic exercise

20
Technical skill:
Shooting from
various angles

20
Game situation:
3 v 2 with
2 goalkeepers

20
Game with
a theme:
5 v 5 with 5
support players

20
Game

Warm up   **15**

2 games of 4 v 4

Set two fields of 15m x 20m
On field A a game of 4 v 4 is played
with goalkeepers.
At the command of the coach the
teams change halves attacking
the other goal.

On field B a 4 v 4 game is
played scoring a goal by
dribbling over the end
zone.
Encourage dribbling and
feints.

Physical   **20**

Athletic exercise
The players runs to cone A or B ,
receives the ball and passes it back
to his teammate. 2:00 for each
series to repeat 4 times.

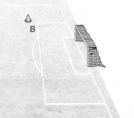

20 mt

(20) ─────────────────────────────── Technical skill

Shooting from various angles

Players run with the ball to the cone and then shoots in goal. After shooting from each of the 4 cones the players change the side and shoot with the other foot.

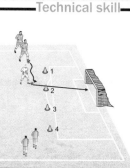

Variation:
Once the player reaches the cone he passes the ball on one side of the cone and he runs on the other side and shoot in goal.

(20) ─────────────────────────────── Game situation

3 v 2 with 2 goalkeepers

Set a field of 30m x 15m
A game of 3 v 2 is played.
The team of 3 players must shoot in goal. If one of the defenders wins the ball he can shoot in the other goal without any pressure.

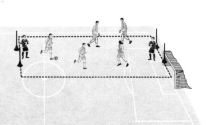

(20) ─────────────────────────────── Game with a theme

5 v 5 with 5 support players playing only headers

5 v 5 game with 5 players on the outside as support who can play only with their head.
The goalkeeper must play the ball to the support players for them to head the ball in the field.
After 6' the teams are changed.

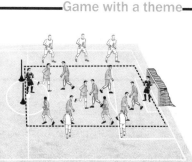

(20) ─────────────────────────────── Game

Objectives

Individual tactics:
Dribbling and feints

Group tactics:
Ball possession and counter attack

Technical training:
Shooting

Physical conditioning:
Quickness

Organization

(105)

(15)
Warm up:
1 v 1 with feints
and dribbling

(10)
Physical:
Triangle

(20)
Technical skill:
Shooting from
various angles

(20)
Game situation:
6 v 6 + 6 support

(20)
Game with
a theme:
11 v 11 ball
possession

(20)
Game

Warm up　　　**(15)**

1 v 1 with feints and dribbling
Set various mini fields of 5m x 10m
with small goals of 1m. Players can
score by dribbling over the end
zone but must dribbling and feint.
Every 2:00 change the players.

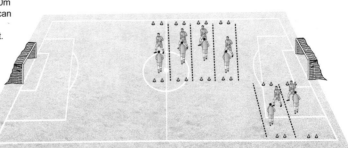

Physical　　　**(10)**

Triangle
Three players form a triangle passes
the ball using volley with inside foot,
volley with instep, half volley inside
foot, heading, chest and thigh.

:30 for each technique.

(20) ──────────────────── Technical skill

Shooting from various angles

Every players dribble to the cone and as shown in the figure, he turns around it and shoot in goal. Continue the exercise for the other 2 cones then change side and shoot with the other foot.

Variation:
Ask the player to shoot on the near or far post.

(20) ──────────────────── Game situation

6 v 6 + 6 support

2 teams play a 6 v 6 game with only two touches. 6 support players are placed two on the sides and 4 nears the goals.
The support players can only play one touch. Two points if a goal is scored after a give and go with the support player.

Variations:
The support players can only play the ball on the ground. The support players can only play the ball in the air.

(20) ──────────────────── Game with a theme

11 v 11 ball possession and counter attack

Two teams play a game. Team A can score only from ball possession building up from the back. Team B can score only within a time limit from the time they win the ball.
After 10:00 change the teams.

Variation:
1) Teams play with limited touches.
2) Team B can only play forward.

(20) ──────────────────── Game

Objectives

Individual tactics:
Dribbling

Group tactics:

Technical training:
Shooting with a volley using the weaker foot

Physical conditioning:
Speed endurance

Organization

110

20
Warm up:
3 stations working
only with weaker
foot

20
Physical:
Athletic exercise

15
Technical skill:
3 stations shooting

15
Game situation:
2 v 2 with
4 goals

20
Game with
a theme:
6 v 6 with 6 goals

20
Game

Warm up — 20

3 stations working only with weaker foot

A) Running with the ball in a straight line and in a diagonal line turning around the cone.

B) After 3 juggles stop the ball with different parts of the foot: sole, instep, inside, outside.

C) Players are in pair and control the ball with the chest, instep volley, inside volley, half volley.

Physical — 20

Athletic exercise

Two teams are inside the circle. At the coach's command the grey team runs out of the circle towards the cone and come back in the circle. In the meantime the blue team must pass the ball inside the circle as many times as possible before the grey team is back. Once all the grey team is back the blue team runs out and the process is repeated.

4 repetitions of 2:00 with a recovery time of 1:30.

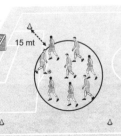

15 mt

15

3 stations shooting

3 exercises to improve shooting with a volley.
a) the player juggles the ball and hit it over two poles.
He runs around the pole and hit the ball on a volley or after one bounce trying to shoot in one of the small goals.
b) A player runs to the cone enters the box and hit the ball that has been thrown by another player with his hands. The player must hit the ball with a volley. c) Player A is positioned as in the figure and performs bicycle kicks or half bicycle kicks. The ball is played in by another player with his hands.

15

2 v 2 with 4 goals

After a quickness circuit the players run into the field to play a 2 v 2. The team that wins the ball can score in the goal called by the coach. If the defending team wins the ball they can score on the opposite goal.

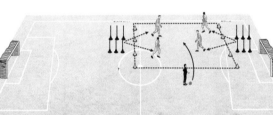

20

6 v 6 with 6 goals

In field of 40m x 30m we play a 6 v 6 games with 6 small goals. The goal is scored dribbling through the goals but a player cannot score on the same goal twice. Players are encouraged to dribble. In the meantime the players not involved performs a series of dribbling and feints on a circuit with final shooting. Every 5:00 a new team is called to play.

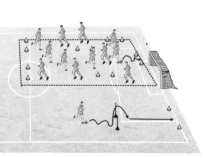

20

Objectives

Individual tactics:
Dribbling and feints

Group tactics:

Technical training:
Shooting and receiving the ball

Physical conditioning:
Psychokinetic and speed endurance

Organization

110

10
Warm up:
Game of 4 colors

20
Physical:
Comparison game

20
Technical skill:
Receiving and shooting

20
Game situation:
1 v 1, dribbling and shooting

20
Game with a theme:
3 goals to dribble through

20
Game

Warm up

15

Game of 4 colors

Divide the players in 2 groups of 8-10 players. Divide each group in 4 colors. The game is played with feet.
a) the player passes the ball and calls the color he passes to
b) the second player passes the ball and calls a color different to the one he passes to
c) the third player passes the ball and calls a color that the receiving player must pass to
d) the fourth player passes the ball and calls a color that the receiving player cannot pass to.

Variation:
Use two balls.

Physical

20

Comparison game

Set the field as in the figure. The grey team shots in goal and try to score as many goals as possible while the blue team performs a series of 20m sprints.
The players of the blue team must perform 4 sprints of 20m each. Player A sprints towards B who starts the sprint as soon as he is touched by A.

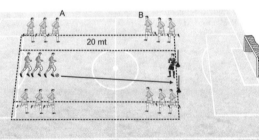

Receiving and shooting

a) the player receives the ball from the coach after making a diagonal run and shooting quickly in goal. The ball is received on the ground, in mid air and in the air.

b) the player receives the ball with his back to the goal and he must receive and control the ball around the cone. The ball is received on the ground, in mid air and in the air.

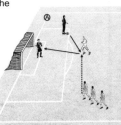

1 v 1, dribbling and shooting

Two players dribble around the poles and at the end they leave the ball and sprint inside the field to win a ball played by the coach. The first player on the ball must dribble to a cone outside the field and turn around, the other player must run and touch a cone beside the goal, and afterward he must quickly apply pressure to the player with the ball who can shoot in goal but only after overtaking the defender with a dribbling move.

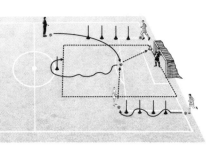

3 goals to dribble through

Set a field of 50m x 35m with 3 goals on each side of the field. Players must dribble through the goals, once they dribble past the goals they pass the ball to the coach and shoot on goal first a stationery ball placed in front of the goal with their weaker foot and then the ball passed back by the coach with their strong foot. On the other side the player only shoot the stationery ball. After 10:00 the teams change the sides.

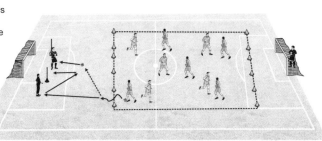

Objectives

Individual tactics:
Dribbling and feints

Group tactics:
Overlapping, crossing and attacking interchange

Technical training:
Shooting

Physical conditioning:
Quickness

Organization

110

20
Warm up:
Dribbling and feints

10
Physical:
Basic technique with three players

15
Technical skill:
Dribbling and feints

25
Game situation:
Collective movements

20
Game with a theme:
6 v 6 game

20
Game

Warm up

20

Dribbling and feints

In a 20m x 20m field players dribble around in pairs with one ball. At he command the player with the ball dribbles towards his teammate who acts as passive defender, and performs various type of feints. Once the player overtake his teammate he passes the ball to him and the process is repeated by the other player.

Variation:
After 10:00 the defenders become active.

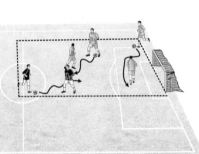

Physical

10

Basic technique with three players

Three players in line perform basic technique of receiving and passing, :30 for each player.

(25) ───────────────── **Technical skill**

Dribbling and feints

Place 2 lines of 5 poles in a zigzag
pattern for a total distance of 7-10m
Dribble around the poles using
various feints: Step over to open
and step over to close, double
step over, Maradona move,
Matthews move, etc.

Variation:
Introduce a time limit to increase the intensity and quickness.

(15) ───────────────── **Game situation**

Collective movements for the overlapping and crossing

Place all the players in their own positions except
the central defenders. The exercise starts with the
ball to one of the center midfielders who passes
to the opposite fullback. The fullback plays
the ball to the winger and overlaps. The
winger receives the ball dribble inside
and passes the ball to the overlapping
fullback who crosses the ball. At
the time of the cross the forwards
must crisscross while one center
defender will move from the
post to mark one of the
forwards.

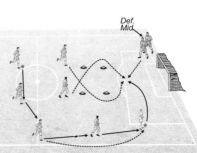

Variation:
Introduce another center defender to create a 2 v 2 in the penalty area.

(20) ───────────────── **Game with a theme**

6 v 6 game

The field is divided on a central zone, two
attacking zones and two lateral zones.
In the central zone we place 6 players
from each team. In the lateral zone one
player from each team and in the
attacking zone only one forward. We
start to play in the central zone
and after a determined number of
passes the ball is played to the
lateral player who can cross
without any pressure from
the other player, in the
attacking zone.

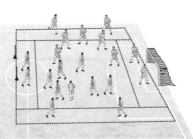

Variations:
Introduce one defender in the attacking zone to create 1 v 1 – change the positions of the players.

(20) ───────────────── **Game**

Objectives

Individual tactics:
  Shielding of the ball

Group tactics:

Technical training:
  Receiving

Physical conditioning:
  Conditional abilities in situation

Organization

100

15
Warm up:
Ball possession

15
Physical:
Athletic exercise

15
Technical skill:
Analytic exercise

15
Game situation:
1 v 1

20
Game with
a theme:
4 v 4

20
Game

Warm up — 15

Ball possession
  Each player with a ball perform
  exercises of juggling, After a few
  juggles the player receives and
  control the ball with inside, outside
  and instep of foot, with the thigh
  and with the head.
  After each receiving of the
  ball the player must run with
  the ball.

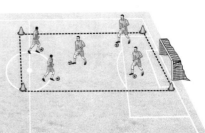

Physical — 15

Athletic exercise
  Specific exercises for the wingers
  and full backs and for the rest of the
  group,
  The wingers and full back spring
  for 20m, they stop to the cone
  and re-start the run towards the
  ball and they cross.
  The rest of the group skip
  over the hurdles performing
  a heading gesture without
  the ball and then try to
  head the ball from the
  cross.

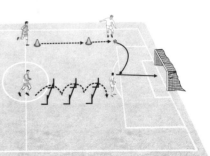

(15) ──────────────────── Technical skill

Analytic exercise

Two teams of 6 players each.
Player A passes to B and takes
his place. Player B performs a
countermovement, receives the
ball and shoot.
Use various type of receiving
the ball and play in a
competition format.

(15) ──────────────────── Game situation

1 v 1

Set various mini-fields with a small
goal and an end zone. The attacker
receives the ball and must shield it
for 5:00 from the pressure of the
defender. If he is successful in
protecting the ball he can shot
in the goal.
If the defender wins the ball
he will score dribbling in
the end zone.

(20) ──────────────────── Game with a theme

4 v 4 with pass to the captain

Set a 20m x 20m field where a 4 v 4
game is played.
The objective of the attacking team
is to pass the ball to the captain
who is standing in a box 5m x
5m at least 5m away from the
field.
To score a point the captain
must receive the ball and
shield it for 5:00 from
the pressure of one
opponent.

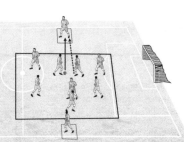

(20) ──────────────────── Game

Objectives

Individual tactics:
Shielding of the ball

Group tactics:

Technical training:
Receiving

Physical conditioning:
Conditional abilities in situation

Organization

110

15
Warm up:
Three teams and
three balls

15
Physical:
Athletic exercise

20
Technical skill:
Passing and
receiving

20
Game situation:
2 v 1 with shielding
of the ball

20
Game with
a theme:
3 v 3 receiving

20
Game

Warm up   15

Three teams and three balls
Three teams move freely around the
field and pass the three balls using
their hands with the only rule to pass
the ball to a different color. Once
the ball is thrown in the air the
player must receive the ball with
his chest and then can catch it
with his hands.

Variations:
a) change the type of
receiving of the ball
b) the whole game is
performed by juggling.

Physical   15

Athletic component

Athletic exercise
The center midfielders are inside the
circle always moving and perform a
"give and go" with the other players
who run with the ball.
Each player run in a 30m circuit
outside the circle with the ball at
maximum speed.

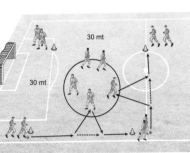

30 mt

30 mt

20 ─────────────────────────── Technical skill

Passing and receiving

As shown in the figure the players are facing each other with a pole in between to act as obstacle. Player A passes to B with the inside of the foot. Player B receives with his left foot and directs the ball to his right foot to pass back to A.

Variation:
• Pass with the right foot and receive with the left foot, -pass with the left foot and receive with the right foot – pass with the right foot and receive with the outside of the left foot and vice versa.

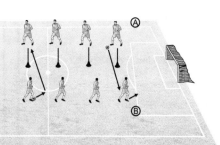

20 ─────────────────────────── Game situation

2 v 1 with shielding of the ball

Set a small field 15m x 10m where the players start in a 1v1 game. Player A passes the ball to B after performing a countermovement and he must shield the ball from the pressure of the defender. After making the pass, player A must run in a figure of eight circuit and at the end of it he can support his teammate to play a 2 v 1 game.

20 ─────────────────────────── Game with a theme

3 v 3 receiving

Play a 3 v 3 in a 40m x 30m field. Outside of the field set 2 fields 5m x 5m where players B and B1 must receive the ball correctly from player A and A1 10 times with both the right and left foot. At the end of the ten receiving of the ball they can join the game. With this game we also train situations of numerical advantage.

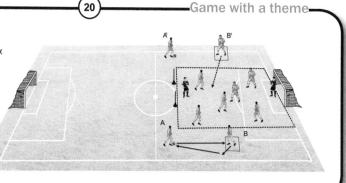

20 ─────────────────────────── Game

Objectives

Individual tactics:
Shielding of the ball

Group tactics:
Movements of center midfielders off the ball

Technical training:
Receiving and shooting

Physical conditioning:
Quickness

Organization

110

15
Warm up:
Defending and attacking

10
Physical aspect:
Technical gesture

20
Technical skill:
Receiving and shooting

20
Game situation:
Movement of the center midfielders

20
Game with a theme:
Different games

20
Game

Warm up　　　**15**

Defending and attacking

Two teams with each player dribbling a ball. At the command each player must defend and shield his own ball and try to kick the opponent's ball out of the field.

Last three players on the field get a point.

The players eliminated wait for the next game by juggling.

Physical　　　**10**

Technical gesture

Players in pairs perform a technical gesture after sprinting towards the cone.
a) volley with the instep
b) volley with inside of the foot
c) half volley.

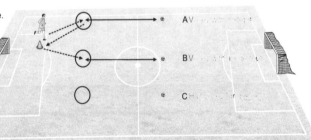

(20) — Technical skill

Receiving and shooting

Players are placed in the three different positions 1 - 2 – 3. Players must attack cone A where they receive the ball from the coach and they first touch it in direction of the goal and then shot in goal.

The three different positions reflect three different angles for receiving the ball.

Variation:
1)Ball passed on the ground (receiving with the inside of the foot and the outside)
2) Ball bouncing 3) Ball in the air and control with the chest

(20) — Game situation

Movements of the center midfielders Without the ball

Based on the 4-4-2 formation place 4 center midfielders on the field. In front of them 4 players who pass the ball to each other. The 4 midfielders focus on movements of closing down and applying pressure to the player in possession of the ball.

The other players not involved in the activity work on juggling the ball with various themes.

Variation:
Change the position of the 4 opponents (in our half or in their half) and change the tactical placement of the 4 opponents.

(20) — Game with a theme

Different games

Two teams are involved in different games. The players of team A stand inside small squares and they must pass and receive correctly inside the box 40 passes in the air.

The players of team B instead are involved in a game of crossing.

The crosser cross inside a box placed in the middle of the penalty box for his teammate.

The attacker must receive inside the square and score in the goal with the color called by the crosser.

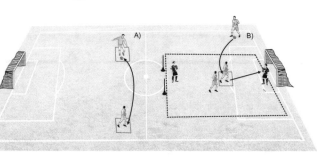

(20) — Game

Objectives

Individual tactics:
Shielding of the ball

Group tactics:

Technical training:
Receiving and passing

Physical conditioning:
Conditional abilities in situation

Organization

110

15	**15**	**20**
Warm up: Passing and receiving	Physical aspect: 1 v 1 sprinting to the ball	Technical skill: Receiving and orienting the ball

20	**20**	**20**
Game situation: 1 v 1 receiving and shielding of the ball	Game with a theme: 8 v 8	Game: 4 v 4 tournament

Warm up   **15**

Passing and receiving with three players

Three players are 15m apart and pass the ball using two touches and focusing on the receiving of the ball, once they pass they change the position.

Variation:
1) ball on the ground
2) ball in the air (increase the distance to 30m)
3) keep the ball up in juggling.

15 mt

Physical   **15**

1 v 1 sprinting to the ball

The two players sprint for 15m to get to the ball first starting from the following positions:
a) laying on the ground on their back
b) laying on the ground on their stomach
c) sitting
d) after jumping on the spot.

 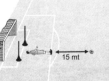

15 mt

(20) ———————————————————— Technical skill ——

Receiving and orienting the ball

Player A stands in the middle of the field with three small goals. Player B passes the ball to A and calls a number. Player A must control the ball towards the goal called by player B. Once he goes through the goal he passes back to B. The player works for 1:20 before switching positions

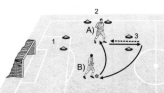

Variations:
1) receive ball on the ground 2) receive ball in mid-air 3) receive ball in the air
4) use all parts of the body to control the ball.

(20) ———————————————————— Game situation ——

1 v 1 receiving and shielding of the ball

The attacking player stands in the center of the field. The coach passes the ball and call the color of a goal to score on. At the same time the defender closes in on the attacker, who must control the ball and orient it towards the goal with the color opposite to the one called by the coach. He must shield the ball for 5:00 and start the 1 v 1 from a position with his back to the opponent.

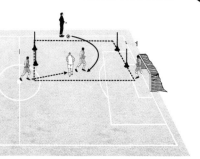

(20) ———————————————————— Game with a theme ——

8 v 8

Set 3 boxes of 2 x 2 on each attacking zone.
The game is an 8 v 8 and after 5 passes the team in possession of the ball can lob the ball in one of the boxes.
If a player receives correctly in the box he can pass the ball to a teammate in front of the goal who can shot without being pressured.

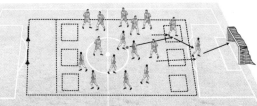

(20) ———————————————————— Game ——

4 v 4 tournament

Objectives

Individual tactics:
Positioning to defend the goal

Group tactics:

Technical training:
Heading

Physical conditioning:
Conditional abilities in situation

Organization

 115

(20)
Warm up:
Throw-in &
heading

(15)
Physical aspect:
Game to team

(20)
Technical skill:
Heading in mini
goals

(20)
Game situation:
3 v 3 defending
the goal

(20)
Game with
a theme:
8 v 4

(20)
Game

Warm up — (20)

Throw-in and heading
Player A throws-in for player B and
positions himself to defend the goal
without using his hands. B heads
the ball in.
The exercise should be done in a
competitive format.

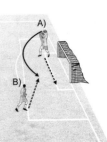

Physical — (15)

Game to team
The blue team must make 4 passes
before long passing to the gray team.
The black team attacks the team in
possession of the ball.
If they win the ball the team losing
possession must apply pressure
to the other team.

(20) ─────────────────────── Technical skill

Heading in mini goals

Two teams with eight players play in two different fields. In each field place two mini goals identified by different colors. Player A performs a throw in and calls a color of one of the two goals. Player B after a slalom around the poles heads the ball towards the goal called by the teammate.

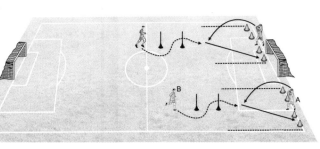

(20) ─────────────────────── Game situation

3 v 3 defending the goal

2 teams of 6 players each.
Three players are on the side to cross the ball the other 3 are in the field ready to shot.
The teams will alternate crossing the ball.
A goal scored with the head counts twice. One player per team can defend the goal without using his hands.

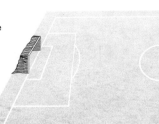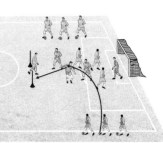

(20) ─────────────────────── Game with a theme

8 v 4

Set a field long and narrow.
The game is 4 v 4 with a goalkeeper in each goal.
The team in possession of the ball has available 4 players in support on the outside.
The goal can be scored only on a header.

(20) ─────────────────────── Game

Objectives

Individual tactics:
Collective movements when in non possession of the ball

Group tactics:

Technical training:
Heading

Physical conditioning:
Quickness

Organization

115

(15)
Warm up:
2 v 2 soccer tennis
tournament

(10)
Physical aspect:
Conditioning circuit

(20)
Technical skill:
Three players
juggling

(20)
Game situation:
11 v 0 shadow
play

(20)
Game with
a theme:
4 v 4 + 4 support
players

(20)
Game

Warm up

(15)

Soccer tennis tournament 2 v 2

Set various mini fields and play soccer
tennis with the following rules:
• The ball can only be slammed with
a header
• Only three touches per team
allowed and the ball can only
bounce once
• The player must play the
ball to his teammate before
sending the ball across.

Physical

(10)

Conditioning circuit

First the players perform a low skip,
then they go under the rope moving
left and right of it and by lowering
and raising their body.
Conclude the circuit by working
inside the circles and sprint
for 10m

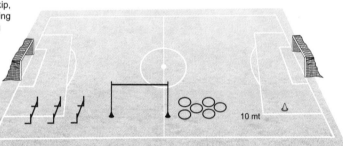

10 mt

(20) ──────────────────────── Technical skill

Three players juggling

Three players work on heading technique, using one or two touches. The players try to keep the ball in the air passing it with their head. Player A passes to B, and B passes to C. Player B then change position with A. Player C passes to B and B returns the ball to C who passes to A.

(20) ──────────────────────── Game situation

11 v 0 shadow play

Set the team according to the system played (i.e. 4-4-2) and place 11 flags with a number.
The coach calls a number and the players move and shift by adjusting the team shape based on the number called.

(20) ──────────────────────── Game with a theme

4 v 4 + 4 support players

A game of 4 v 4 is played in a field as shown the figure.
Each team has 4 support players available, 2 in the end zones and 2 on the sides. The support players can pass the ball to each other as well. The game is played first with the hands, then hands and feet and finally only with feet.
The goals scored with a header count two.

(20) ──────────────────────── Game

Objectives

Individual tactics:
Positioning to defend the goal

Group tactics:

Technical training:
Heading

Physical conditioning:
Conditional abilities in situation

Organization

105

15 Warm up: Defending the goal

15 Physical aspect: Motor exercise

20 Technical skill: Four players juggling

15 Game situation: 2 v 1 with heading wall pass

20 Game with a theme: Game of four colors

20 Game

Warm up — 15

Defending the goal

As shown in the figure, player A dribbles the ball to the pole turns around it and shoot on goal.
Player B runs around the pall and proceed toward the goal to defend it.
The team that scores more goals wins.

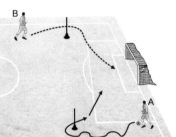

Physical — 15

Motor exercise

The group is divided in three teams. Two teams play a possession game, while the third team performs strengthening exercises (squats, push-ups, sit-ups).
After 10 repetitions the team performing the strengthening exercises tries to gain possession of the ball.

(20)

Four players juggling

Four players stands in line. B passes
to A, A passes back to B. B passes
to D. D passes to C, and C passes
back to D. D passes to A who
in the meantime has changed
position with B.
All this is performed using
the head.

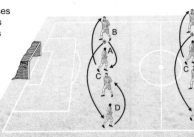
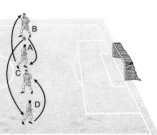

(20)

2 v 1 with heading wall pass

Set a mini field with one regular goal on one end
and two small goals on the other end. Players
A starts from outside the field and must lob
a long ball towards players B and C. These
players must first perform an athletic exercise
(summersault and slalom through the poles)
then try to attack the ball and head it back
to player A. The player who arrives first
on the ball and heads it back to A plays
with him a 2 v 1 trying to score on the
large goal. The other player defends
the goal helping the goalkeeper and
if he gains possession of the ball
he can score in one of the two
small goals.

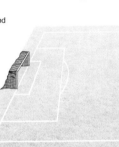
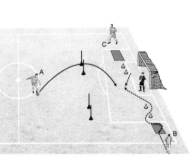

(20)

Game of four colors

The game is 6 v 6. Each team is made
up of two colors.
White players play with their hands
and the black players play with
their heads.
The same thing applies to the
blue team that uses its hands
and the grey team that uses
his head.
The goal can only be
scored with a header.

(20)

Objectives

Individual tactics:
Positioning to defend the goal

Group tactics:

Technical training:
Heading

Physical conditioning:
Conditional abilities in a situation

Organization

110

(20)
Warm up:
Defending the goal

(15)
Physical aspect:
Motor exercise

(15)
Technical skill:
Score by heading

(20)
Game situation:
2 v 2, score and
defend goal

(20)
Game with
a theme:
6 v 6 with
goals

(20)
Game

Warm up — (20)

Defending the goal
Four players stand on the sides of the
box and pass the ball to each other
with the objective to score in one
of the mini 4 goals placed in the
center.
The defender must position
himself to defend the goals
based on the lateral
passes.

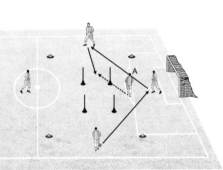

Physical — (15)

Motor exercise
3 players with different color jumps
over three obstacles and sprint 15m
towards the ball.
The first player on the ball calls a
color to play a 2 v 1 against the
other color.
The goal is scored by dribbling
through the goal.

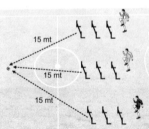

15 mt

15 mt

15 mt

15

Score by heading

In a field of 20m x 20m we place 4 goals.
At each corner there are 4 players that cross the ball inside the field where 4 players will try to head the ball in one of the goals.

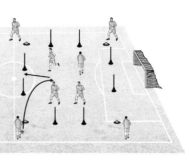

20

2 v 2, score and defend the goal

In a field of 15m x 15m we place 4 goals. At each corner there are 4 players with one ball each, two balls per team.
At the coach's command the players on the corner will cross the ball to their teammates, resulting in a 2 v 2 game and the goal must be scored with a header.

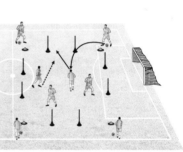

20

6 v 6 with 6 goals and shooting from outside the penalty area

Set the field as shown in the figure.
The game is played 6 v 6. The players can only shoot from the center zone.
The defending team must defend all the three goals.

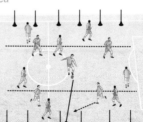

20

Objectives

Individual tactics:
Positioning to defend the goal

Group tactics:
Collective movements when in possession
of the ball

Technical training:
Heading

Physical conditioning:
Quickness

Organization

110

10
Warm up:
Crisscross

10
Physical aspect:
Motor exercise

20
Technical skill:
Attacking &
defending

20
Game situation:
Collective tactics

20
Game with
a theme:
6 v 6 with 6 outside
supporting players

20
Game

Warm up   **10**

Crisscross, jump and head the ball
Two teams start on the side of the
goals.
Player A performs a throw-in for
player B who jumps over the
obstacles and head the ball trying
to score.

Physical   **10**

Motor exercise
Players run in slalom without the ball
then perform a technical gesture with
their team mate.

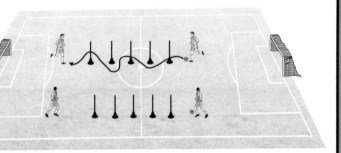

(20) — Technical skill

Attacking and defending

Player A has the objective to score by shooting.

Player B must obscure the shooting path and try to intercept the shot.

Both players are separated by a line and cannot get into contact.

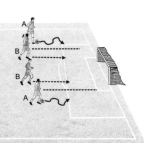

(20) — Game situation

Collective tactics

The coach set the team according to its playing style in a smaller field. Players must perform movements appropriate when in possession of the ball from the build up to the finishing in goal.

(20) — Game with a theme

6 v 6 with 6 outside support players

Players can only play with their heads and can use the outside support players.
When the ball hits the ground possession change hands.

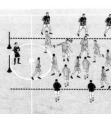

(20) — Game

Additional Drills For Strength & Power

Jumping & technical gesture

Obstacles at approximately 2m distance. The player must jump over the obstacles and perform a technical gesture in between the obstacles with his teammates   (i.e. inside of the foot volley, inside of the foot on the ground, instep volley, half volley).

Heading

Player A runs around the cone, climbs on the block, performs a ply metric jump and head the ball crossed by the other player.

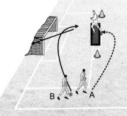

Juggling in pairs

Players juggle the ball and after they pass the ball to the other player they sit on the ground and get up without the use of their hands.

Set 4 stations where players perform different exercises.
a) Heading
b) Starting from laying on the ground the player gets up and perform the technical gesture
c) The player is laying on the ground on his stomach, he gets up and perform the technical gesture.
d) From a sitting position the

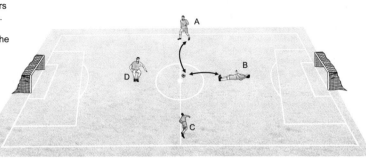

player gets up without using his hands and performs the technical gesture.
Every exercise must be repeated three times.

Two players on the side of the field perform various strengthening exercises (squats, push-ups, sit-ups). Each exercise must be repeated 8-10 times and afterwards the players get to play in a 1v1 game trying to score in the goal with a goalkeeper.

Station A – push ups
Station B – sit ups
Station C – Squats
Station D – heading the ball with the teammate who must repeat the same exercises

Players perform 8 to 10 repetitions per station.

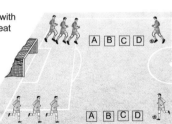

Additional Drills For Endurance

Ball possession

The coach calls the color of the team whose players must run around the field three times at high speed. At the same time the two teams inside the field will play a game of possession. The team that has made the most passes wins the game.
The exercises duration is 5:00 with 2:00 recovery.

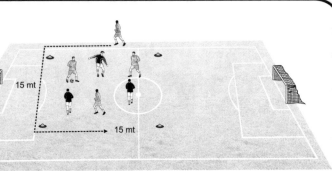

15 mt

15 mt

Running with the ball

Players dribble around the field at a relaxed speed. Players get in the field and dribble at maximum speed towards the cones with the color called by the coach. The coach will call the colors in a fast succession.
Duration 1:00.

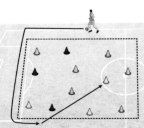

High intensity 3 v 3

In a small field players play 3 v 3 keeping the ball always moving and in play.

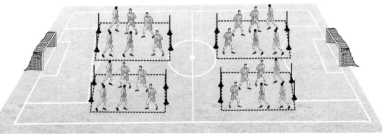

Steal the ball

Player A must try to kick the ball of the other players out of the field.
The field must be wider.

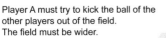
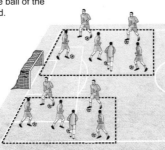

Keep away with colors

Players wear two colors and form a circle.
Two players in the middle. Player grey can only attack blue players and the blue player can only attack the grey players.

Use a wide circle. Players in the middle work for 1:00

1 v 1

Players wear different colors and are assigned a number of 1 or 2. The players run around the field.
When the coach calls the color and the number the players run in the small field to play a 1 v 1.
A the end of the game they go back in the group.

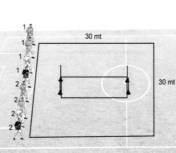

Additional Drills For Speed With The Ball

Speed of reaction

The game is played in a smaller field.
At the whistle the teams change direction of attack and attack the opposite goal.
The interval between a whistle and the other varies.

Speed of reaction in pairs

Player A dribbles the ball in the 30m space and perform change of pace and direction at his own liking.
Player B must perform the same movements.

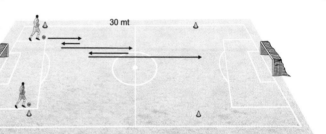

Speed of the reaction movement

Player sprints with the ball for 15m then the coach calls a color and the player must sprint another 10m to the color.

Variations:
a) Change the sequence of the colors
b) Call two colors where the players cannot sprint to

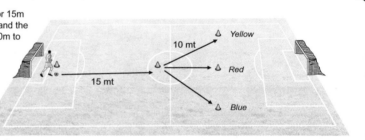

Speed of reaction & 1 v 1

Player A and B sprint at the maximum speed and at the whistle of the coach they enter the field to play in a 1 v 1.

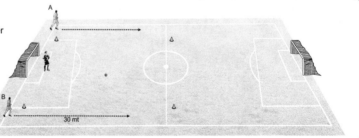

Acceleration 2 v 1

Starting from a standing position 2 attackers and 1 defender sprint and play a 2 v 1 game.
The attackers start 15m away from the ball and the defender 20m away.

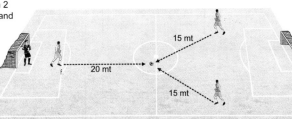

Accelerations & decelerations

Starting from a standing position the player sprints from A to B at the maximum speed. Between B and C the player decelerates and performs the technical gesture. The player sprints again from C to D and perform more technical gestures between D and E. The exercises is ended with final sprint from E to F.

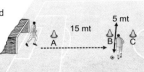

Acceleration & deceleration In 1 v 1

Players A and B sprint at the maximum speed.
At the coach's whistle they stop, walk for 3 steps and sprint again towards the ball in the field.
After reaching the ball they play in a 1v1 game.

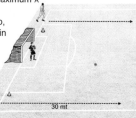

Acceleration, deceleration & speed of reaction

Players run with the ball over a 50m distance at maximum speed. At the coach's whistle they stop and start again in a more relaxed run. At the next whistle they start sprinting again at the maximum speed. The whistles are repeated for the whole length of the run setting the tempo.

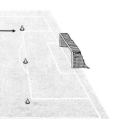

Weekly block 1	Weekly block 2	Weekly block 3	Weekly block 4
Lesson 1 / Lesson 2 / Lesson 3	Lesson 1 / Lesson 2 / Lesson 3	Lesson 4 / Lesson 5 / Lesson 6	Lesson 4 / Lesson 5 / Lesson 6

september

Weekly block 5	Weekly block 6	Weekly block 7	Weekly block 8
Lesson 7 / Lesson 8 / Lesson 9	Lesson 7 / Lesson 8 / Lesson 9	Lesson 10 / Lesson 11 / Lesson 12	Lesson 10 / Lesson 11 / Lesson 12

october

Weekly block 9	Weekly block 10	Weekly block 11	Weekly block 12
Lesson 13 / Lesson 14 / Lesson 15	Lesson 13 / Lesson 14 / Lesson 15	Lesson 16 / Lesson 17 / Lesson 18	Lesson 16 / Lesson 17 / Lesson 18

november

Weekly block 13	Weekly block 14
Lesson 19 / Lesson 20 / Lesson 21	(empty)

december

Weekly block 15	Weekly block 16
Lesson 22 / Lesson 23 / Lesson 24	Lesson 22 / Lesson 23 / Lesson 24

january

Weekly block 17	Weekly block 18	Weekly block 19	Weekly block 20
Lesson 25 / Lesson 26 / Lesson 27	Lesson 25 / Lesson 26 / Lesson 27	Lesson 28 / Lesson 29 / Lesson 30	Lesson 31 / Lesson 32 / Lesson 33

february

Weekly block 21	Weekly block 22	Weekly block 23	Weekly block 24
Lesson 31 / Lesson 32 / Lesson 33	Lesson 31 / Lesson 32 / Lesson 33	Lesson 34 / Lesson 35 / Lesson 36	Lesson 34 / Lesson 35 / Lesson 36

march

Weekly block 25	Weekly block 26	Weekly block 27	Weekly block 28
Lesson 37 / Lesson 38 / Lesson 39	Lesson 37 / Lesson 38 / Lesson 39	Lesson 40 / Lesson 41 / Lesson 42	Lesson 40 / Lesson 41 / Lesson 42

april

Weekly block 29	Weekly block 30
Lesson 43 / Lesson 44 / Lesson 45	Lesson 43 / Lesson 44 / Lesson 45

may

Weekly block 1				**Weekly block 2**				**Weekly block 3**				**Weekly block 4**		
Lesson 1	Lesson 2	Lesson 3		Lesson 1	Lesson 2	Lesson 3		Lesson 4	Lesson 5	Lesson 6		Lesson 4	Lesson 5	Lesson 6

september

Weekly block 5				**Weekly block 6**				**Weekly block 7**				**Weekly block 8**		
Lesson 7	Lesson 8	Lesson 9		Lesson 7	Lesson 8	Lesson 9		Lesson 10	Lesson 11	Lesson 12		Lesson 10	Lesson 11	Lesson 12

october

Weekly block 9				**Weekly block 10**				**Weekly block 11**				**Weekly block 12**		
Lesson 13	Lesson 14	Lesson 15		Lesson 13	Lesson 14	Lesson 15		Lesson 16	Lesson 17	Lesson 18		Lesson 16	Lesson 17	Lesson 18

november

Weekly block 13				**Weekly block 14**		
Lesson 19	Lesson 20	Lesson 21				

december

Weekly block 15				**Weekly block 16**		
Lesson 22	Lesson 23	Lesson 24		Lesson 22	Lesson 23	Lesson 24

january

Weekly block 17				**Weekly block 18**				**Weekly block 19**				**Weekly block 20**		
Lesson 25	Lesson 26	Lesson 27		Lesson 25	Lesson 26	Lesson 27		Lesson 28	Lesson 29	Lesson 30		Lesson 31	Lesson 32	Lesson 33

february

Weekly block 21				**Weekly block 22**				**Weekly block 23**				**Weekly block 24**		
Lesson 31	Lesson 32	Lesson 33		Lesson 31	Lesson 32	Lesson 33		Lesson 34	Lesson 35	Lesson 36		Lesson 34	Lesson 35	Lesson 36

march

Weekly block 25				**Weekly block 26**				**Weekly block 27**				**Weekly block 28**		
Lesson 37	Lesson 38	Lesson 39		Lesson 37	Lesson 38	Lesson 39		Lesson 40	Lesson 41	Lesson 42		Lesson 40	Lesson 41	Lesson 42

april

Weekly block 29				**Weekly block 30**		
Lesson 43	Lesson 44	Lesson 45		Lesson 43	Lesson 44	Lesson 45

may

Annual Training Schedule

september

october

november

december

january

february

march

april

may

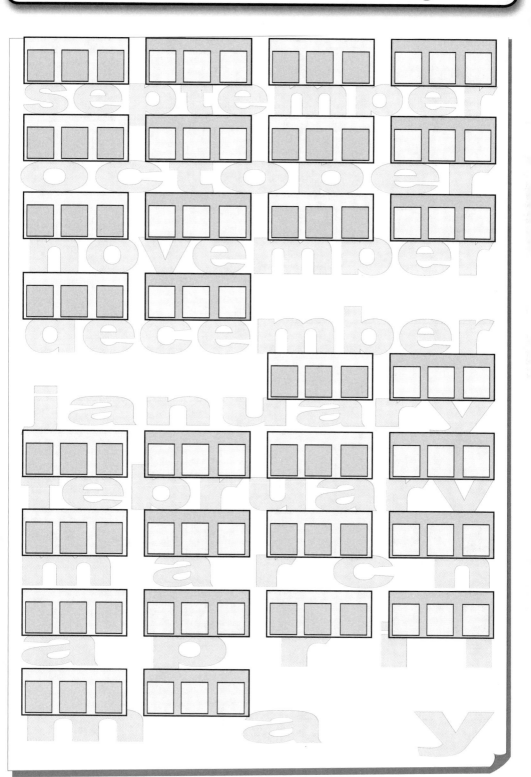

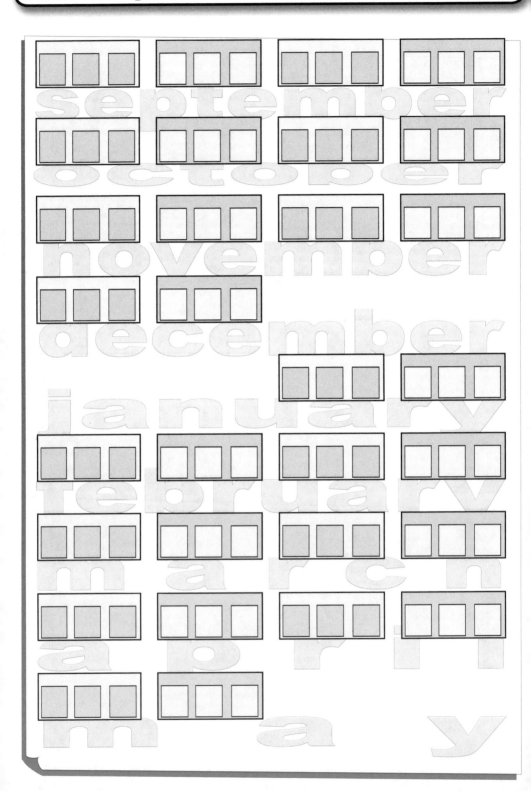

MIRKO MAZZANTINI
Born in Fucecchio on 25/04/77
Degree in Economy and Management
UEFA level I certified
Head Coach of u16 Empoli FC
Staff coach of Empoli Fc for 10 years

SIMONE BOMBARDIERI
Born in Empoli on 31/12/1976
Degree in Motor Science
Uefa level I certified
Empoli FC Soccer School instructor
Staff coach of Empoli FC for 9 years
Head Coach of u13 Empoli FC

TOMMASO TANINI
Born in Firenze on 11/10/1978
Uefa level I certified
Staff coach of Empoli FC for 6 years
Fitness and athletic trainer of u14 Empoli FC
Empoli FC Soccer School Instructor

FAUSTO GARCEA
Born in Paola (Cosenza) on 13/09/1961
Degree in Motor Science
Professor in Motor Sciences at the University of Firenze and sport
team methodology at the Universities of Siena Pisa and Firenze
Uefa level I certified, fitness and athletic trainer
Director of Empoli FC Soccer School since 2005/2006.

ANDREA AGNOLONI
Born in Firenze on 29/04/1962
Canada Soccer Association B Licence Certified
Staff Coach of North Shore Soccer
Development Centre in North Vancouver, BC Canada
Head Coach of u14 Mountain WFC (USL Y League)
Goalkeeper instructor

More Soccer Italian Style Titles Available...

• **Soccer Training Sessions u13 & u12 Book**

• **Ball Control DVD & Blu-ray**

• **Ball Control 2 DVD & Blu-ray**

• **Individual Defending DVD & Blu-ray**

SoccerCoaching.com/ItalianStyle

PO Box 5550
Pleasanton CA 94566 USA
1.800.762.2376
SoccerCoaching.com